STUDIES

ON

SOME FUNGAL BIODETERIOGENS

STUDIES
ON
SOME FUNGAL BIODETERIOGENS

Naresh R. Shah
Senior Lecturer
Department of Museology
Faculty of Fine Arts
M. S. University of Baroda, Vadodara

Arun Arya
Reader
Department of Botany
Faculty of Science
M. S. University of Baroda, Vadodara

BKP

BHARATIYA KALA PRAKASHAN
DELHI (INDIA)

First Edition : 2000

Published by
C.P.Gautam
BHARATIYA KALA PRAKASHAN
3421-A, Narang Colony, Tri Nagar,
Delhi-110035. Phone : 7163083

ISBN : 81-86050-19-1

Laser Type Setting by
A-ONEGRAPHICS
X-4, Gali-2, Brahmapuri, Delhi-110053
Phone : 218 3470

Printed at Santosh Offset Press
40, Anand Nagar, Inderlok, Delhi
Phones : 511 6763, 511 7390

Dedicated to my beloved Mother.........

– Naresh R. Shah

Dedicated to all those concerned with conservation of cultural heritage

– Arun Arya

Foreword

I feel a genuine pleasure in introducing the book "Studies on Some Fungal Biodeteriogens" written by two enthusiastic researchers Dr. Shah and Dr. Arya. The book, without any hesitation, will be useful to mycologists working on biodeteriogens and museologists whose job it is to protect the preserved articles, some of which represent important cultural heritage.

Various museum articles, wooden and fiber, are facing serious threat from fungi and insects. Unfortunately, there is no proper understanding of their cause, and prevention and control. The book provides the much needed information on 'science' and 'art' related to biodeterioration.

The book describes in a very 'easy to understand' style the commonly encountered fungi, their isolation techniques and identification. There is in-depth study of their biochemical potentials. There are exclusive chapters on aeromycoflora and on prevention & control measures which will be highly useful to museum personnel.

The book is a very timely publication providing plethora of information of fungal biodeteriogens which is sure to encourage more work in this area.

Dr. H.C. Dube
F.N.A.Sc.
Professor & Head
Dept. of Life Sciences
Bhavnagar University
Bhavnagar - 364002.

Preface

The present work entitled **Studies on some Fungal Biodeteriogens,** is a detailed study of certain fungi causing severe damage to museum materials. It is true that owing to the passage and ravages of time the **cultural property** disintegrates. But there are several other agencies too taking part in disintegration. Organic materials are much more susceptible to deterioration than the inorganic ones. Numerous agents cause damage to cultural property but among these biological agents like fungi cause havoc in favourable environmental conditions. **Conservation** is one of the five prime pillars of Museology. Being a museologist we are interested in conserving the cultural objects. Microbes like fungi have attracted our attention since long. The most part of the text is an outcome of Ph.D. programme undertaken by N.R.Shah. The studies are presented in the book in such a form that it will help the scientists and Museum personnels to protect the valuable cultural property from fungal deterioration.

In the present study, more emphasis has been laid to study the behaviour of the selected fungi which includes Cultural, Biochemical, and Control Studies. To answer some questions like, why some fungi occur in certain environments? How they are maintained? And how they have adapted to prevailing conditions? Sufficient information on biochemical and physiological behaviour of fungi is required. An attempt has been made to find out different fungi associated with biodeterioration of various museum objects. It includes an aerobiological survey of museums and libraries. The correlation between the fungi present in the atmosphere and the same fungus affecting the organic objects has been established. The degradation of different woods and papers is also studied which gives an idea about the degrading potential of different fungi. At the end efforts are made to find out suitable control measures. These results will be surely useful to understand the biology of biodeteriogens and minimizing the Fungal problems in Museums.

It gives us immense pleasure to extend our deep sense of gratitude to Prof. R.T.Parikh, Retired Head of the Department of Museology, M.S.University of Baroda for his encouragement, inspiration, suggestions and valuable guidance throughout the present investigation. We are thankful to Shri R. P. Shah for his constant encouragement and to Shri Sunjay Jain, Offg. Head, Department of Museology for sparing his valuable time to go through the manuscript. Thanks are also due to Shri Kirit Rehevar and Dr. Ramana for their family help. Assistance received from Shri M.C. Goswami in preparing histograms and drawings for the book and Shri Viquar Sayed in photography is really praiseworthy. Our thanks are also extended to Shri M.V.Rathwa, Shri Prakash Maske and Shri Santosh Kaleker for their valuable help in experimental work.

We are really indebted to Dr. O.P.Agrawal, Director General, INTACH, Lucknow; Dr. S.M.Nair, Sr. Advisor to WWF and Former Director of National Museum of Natural History, New Delhi; Dr.(Ms) Shashi Dhawan, Head of Biodeterioration Section, N.R.L.C., Lucknow; Dr. S.K. Bhowmik, Former Director, Shri A.R. Shah, Sr. Chemist and Mrs. Indira Mafat, Librarian, Museum and Picture Gallery, Baroda for their help and cooperation. Finally gratitude is expressed to the entire staff of Bhartiya Kala Prakashan Delhi and specially to Shri C.P.Gautam for his personal interest in publishing the book in a nice way.

With great pleasure we express our profound appreciation to our wives Mrs. Daksha N.Shah and Dr.(Mrs.) Chitra Arya for their patience and constant cooperation in this publication. They also inspired us in proceeding from thoughts to print.

Helpful suggestions for further improvements are welcome.

Baroda.
June 1st 1999

N. R. Shah
Arun Arya

Contents

Page No.

Foreword (vii)
Preface (ix)
Explanation of Plates (xiii)

1. **Introduction** 1
   1.1 Cultural Heritage
   1.2 Deterioration
   1.3 Biodeterioration
   1.4 Types of Biodeterioration
   1.5 Biodeterioration of monuments
   1.6 Fungi Affecting Ajanta and Ellora Caves
   1.7 Fungi as Biodeteriogens in Museum
   1.8 Fungi under Study
   1.9 Cultural and Control Studies
2. **Biodeteriogens** 15
   2.1 Different Biodeteriogens
   2.2 Fungi as Biodeteriogens
   2.3 Characters of common Fungal bioderteriogens
   2.4 Morphological Characters of selected Fungi
3. **Investigation techniques** 31
   3.1 Preparation of Culture media
   3.2 Isolation Techniques
   3.3 Study of Aeromycoflora
   3.4 Purification of Fungal Cultures
   3.5 Identification of Fungi
   3.6 Assessment of Fungal Growth and Biochemical Studies
   3.7 Biodeterioration of Wood and Paper
   3.8 Control of Biodeteriogens
4. **Isolation Studies** 41

Page No.

5. Cultural Studies **47**
5.1 Selection of suitable culture medium
5.2 Influence of different hydrogen ion concentrations
5.3 Influence of different temperatures
5.4 Influence of different carbon sources
5.5 Utilization of various carbohydrates and their mixtures
5.6 Effect of different nitrogen sources
5.7 Effect of different sulphur sources
5.8 Effect of different phosphorus sources
6. Aeromycoflora & Biodeterioration Studies **127**
6.1 Studies on indoor Aeromycoflora
- Indoor aeromycoflora of Museum and Picture Gallery
- Indoor aeromycoflora of Dept. of Museology
- Indoor aeromycoflora of Oriental Institute
- Indoor aeromycoflora of Smt. H.M.Library
6.2 Biodeterioration of different woods
6.3 Biodeterioration of different papers
7. Control Studies **144**
7.1 Preventive Measures
7.2 Remedial Measures
a. Fungicides
b. Chemical Control
c. Natural Plant products as fungicides
I. Oil & plant products
II. Aqueous leaf extracts
8. Conclusion **159**
9. Summary **163**
References **166**
Appendices **182**
Index **199**

Explanation of Plates

Plate – I :

A. Photograph of the Flute, a musical instrument made up of bamboo, preserved in the Dept. of Museology, M.S. University, Vadodara.

B. Pavri, an important musical instrument used by different tribals. The black coloured substance as well as fruit body harboured the fungus.

C. Wooden Chinese screen displayed in Museum and Picture Gallery, Baroda. It is decorated with precious stones.

D. Fungus affected pages of an illustrated Parsi manuscript book displayed in the Museum of the Dept. of Museology.

Plate – II :

A. Conidia of *Cochliobolus verrucuosus* (*Curvularia* state) obtained from Flute. X 126.

B. Conidiophore and conidia of *Penicillium citrinum* obtained from Pavri. X 126.

C. Conidia of *Aspergillus ustus* causing damage to wooden Chinese screen. X 126.

D. Perithecium of *Chaetomium cymbiforme* (Note the coiled hairs on perithecium) causing deterioration to the Parsi manuscript book. X 94.

E. Enlarged ascospores of *Chaetomium cymbiforme.* X 126.

Plate-III :

A. Photograph showing Museum and Picture Gallery, Baroda situated in Sayaji Park, Baroda.

B. Indian Art Gallery of Museum and Picture Gallery, Baroda where Indoor aeromycoflora was studied.

C. Inner view of Ethnology gallery of Museum and Picture Gallery, Baroda where indoor aeromycoflora was studied.

Plate-IV :

A. External view of the Department of Museology, M. S. University, Baroda.

B. Inside the Museum of the Department.

Plate-V :

A. Outer view of Smt. Hansa Mehta Library, M. S. University, Baroda.

B. Books kept inside the library.

C. Inside view of one of the library rooms.

Plate-VI :

A. Petridishes showing fungal colonies in the experiment of indoor aeromycoflora of Museum and Picture Gallery, Baroda.

B. Petridishes showing fungal flora of the Museum of the Dept. of Museology, M. S. University, Baroda.

C. Colonies of fungi isolated from the air of Smt. Hansa Mehta Library, M. S. University, Baroda.

Plate-VII :

A. Induced growth of *Cochliobolus verruculosus* on the painting made on card board.

B. Induced growth of *Penicillium citrinum* on the painting made on card board.

C. Induced growth of *Aspergillus ustus* on the painting made on card board.

Plate -VIII :

A. Experiments on deterioration of Teak Wood in culture medium by *C. verruculosus* and *P. citrinum.*

B. Experiments on deterioration of Deodar wood in culture medium by *C. verruculosus* and *P. citrinum.*

C. Erlenmeyer flasks showing deterioration of filter paper *in vitro* by *C. cymbiforme.*

D. Erlenmeyer flasks showing deterioration of common paper *in vitro* by *C. cymbiforme.*

Plate-IX :

A. Effect of 0.5% Eucalyptus oil on *C. verruculosus* in control experiment.

B. Effect of 0.5% Eucalyptus oil on *P. citrinum* in control experiment.

C. Effect of 0.5% Eucalyptus oil on *A. ustus* in control experiment.

Chapter—1

Introduction

"A good museum attracts, entertain, arouses curiosity, leads to questioning and this promotes learning. It is an educational institution that is set up and kept in motion that it may help the members of the community to become happier, wiser and more effective human beings."

(Dana, 1909) Foster, 1986

1.1 Cultural heritage

Today Museums are becoming centres of non-formal education. They impart knowledge to all categories of visitors on different aspects of life, about past and present. The concept of museum as a store house has undergone tremendous change. "The museum today is not merely a repository of objects, it is concerned with the acquisition of the object as an integral part of a specific programme of scientific research, education, conservation, documentation and presentation of the national and cultural heritage" (Baxi and Dwivedi, 1973).

The word **museum** is derived from the Greek 'Mouseion' which was a place of contemplation, a philosophical institution or a temple of Muses. The Romans restricted the meaning of the word museum to a place of philosophical discussion. Since last 18th cen. A.D. a museum has been considered as a building used for the storage and exhibition of historic and natural objects (Lewis, 1986). ICOM (1990) defines museum as a non-profit making, permanent institution in the service of the society and of its developments, communicates and exhibits for purpose of study, education and enjoyment, material evidences of people and their environment.

It gives a glimpse of various human activities. Museum speaks a language of objects. "In the contemporary world, a museum tends to be both an expression of the community and instrument to serve it," (Venugopal & Menon

1995). Hooper-Greenhil (1991) commented, "a museum presents probably the only effectual means of educating the adult, who cannot be expected to go to school like the youth..."

Museum's primary function is to be a source of inspiration because the 'Muses' whose house it is, are goddesses of liberal art and are eternal inspiration of men. A museum should be able to give its visitor a lasting impression, some experience which could introduce a new element into his life.

Museum collection comprising of materials of cultural, artistic, historical and scientific significance from different sources serves as an index of civilization between the present and the past. India is one of those few nations which have a very rich ancient culture. Its several thousand years old cultural history shows a strong thread of continuity. India is still not fully aware of the extent and vastness of cultural property that it owns or, of its antiquarian wealth. Much of it is hidden in temples, palaces and old houses; and a still greater number is buried under ground.

The cultural heritage is perishable. It does not regenerate itself like natural heritage. Owing to passage and ravages of time, it disintegrates. The damage caused to it by human agency is much more than natural forces. Further if left to themselves, even the big monuments will be targets of vandalism. In fact, that is what has been happening. "We know that someone broke the Asokan Pillar at Sanchi to use it as a sugarcane crusher and most of the Bharhut stupa sculptures were used as a building material in the neighbouring villages. Certain other museum objects like textiles, paintings etc. will be eaten up by insects if left to themselves," (Baxi and Dwivedi, 1973). In the words of Coremans (1960), "There is inexorable law which spare no work, past or present; all living matter is subject to decay and one day must perish; like doctors we can only do our best to delay the deadly process."

Several steps have been taken from time to time by Government and Non-Governmental agencies for protecting cultural heritage from deterioration. Many antiquarian laws have been passed by the Indian Government from time to time. Principal among these are the Indian Treasure Trove Act 1878, Ancient Monuments and Preservation Act of 1904 followed by Ancient Monuments and Archaeological site and Remains Act 1958; and later on Antiquities and Art Treasures Act, 1972 (Sarkar, 1981). These antiquarian laws helped a lot in protecting cultural property from disappearance by various means.

Museums too have played a vital role in this direction. One of the basic purposes of collecting the museum objects is to preserve the collection for present and future generations. In England the first public museum was the Ashmolean at Oxford opened in 1683. In France the first museum was Abbey of St. Vincent at Besancon in 1694. The British Museum was established in 1759. In India, Asiatic Society of Bengal founded in 1784, decided in 1796 to display many curiosities it had accumulated, in a suitable place in Calcutta. Thus, society established a museum in 1814. Dr. Wallich, a Danish botanist, was appointed as its first Hon. Curator. It was the beginning of the Indian museum movement.

Placing the objects in a museum is not the end point of preservation. There are number of factors-physical, chemical and biological that need to be controlled in order to protect cultural patrimony from deterioration.

1.2 Deterioration

Deterioration is defined as the alteration in an object produced by interaction between the object and factors of destruction (Agrawal, 1993). **Biodeterioration** is a term that includes all types of damage caused by living entities like insects, fungi, bacteria and actinomycetes making the material unsuitable for its intended use. Nair (1972) defined biodeterioration as "any undesirable change in the properties of the materials caused by vital activities of organisms." The agent causing biodeterioration is known as **biodeteriogen**. In contrast to biodeterioration, which is a natural phenomenon caused by biodeteriogens, the biodegradation is a planned, purposeful breakdown or change caused by microorganisms. In other words biodegradation is the harnessing, by man of the decay abilities of organisms to render a waste material more useful or acceptable. Biodeterioration has direct link with our economy as it is related to the survival of object material and to the maintenance of their quality. An increased awareness of the importance of cultural heritage and the need to protect it from various damaging factors, including biological agencies have brought the study of the protection of museum materials into limelight.

The cyanobacteria can be considered the pioneering inhabitants of a stone surface. They can develop with just carbon dioxide, nitrogen and a trace of salt minerals. Just like certain bacteria they are the other prokaryotes capable to fix the nitrogen from atmosphere. The cell has a polysaccharide sheath that allows both fast absorption and slow release of the humidity. Moreover, this sheath facilitates cell/surface adhesion and overcoming of long dry periods.

Several fungal metabolic products are strongly coloured and the phenomenon of staining of damaged areas on monuments can be the consequence of fungal development. Dark spots are attributed to the presence of fungi from family Dematiaceae of Deuteromycotina, which contain melanin pigments inside the mycelium.

Warping of wood, flaking of paint, corrosion in metals, adhesion of loaded papers, fading of colours, weakening, tendering of organic materials, blackening of lead pigments are results of deterioration, whereas, staining, discolouration, erosion, disfigurement, pitting, tunneling, powder formation etc. are examples of biodeterioration.

Deterioration of museum objects depends upon the nature of material of the object and the environment to which it is exposed. It is observed that objects of organic nature are more susceptible to deteriorating agencies compared to the objects of inorganic nature. There are several agents such as, temperature, relative humidity (RH), light, pollution and biotic agents present in the atmosphere. Biological agents are more devastating towards organic objects. Agrawal (1991) narrates his own experiences in these words, ' I have myself seen a whole book-shelf in the library of museum, totally destroyed by termites in a single night'. According to him 'the books were in a perfect condition in the previous evening. In the morning a mud channel was noticed and the damage was found inside'. In another case, in a palace in Tehran the large sized wooden pillars used for supporting the roof became hollow, on account of the damage by termites, although from out side the carvings, as well as the pillars were intact. It is estimated that if timely discovery would not have been made the entire palace could have collapsed. Equally devastating and quick can be the action of fungi on art objects (Agrawal, 1991).

1.3 Biodeterioration

The biodeterioration of materials is one of mankind's oldest problems and its record can be traced back to biblical times or probably earlier. The estimated cost of repairing damage caused by biodeterioration of timber in buildings in the United Kingdom amounts to 400 million pounds a year and about 150 million of it is spent on toxic chemicals.

Biodeterioration of materials occurs throughout the world, specially in the zones where the environmental conditions permit the growth of organisms. For this reason the problems of biodeterioration are found more severely in the tropical countries like Indonesia, Thailand, Myanmar, Srilanka and India, for the large part of the year.

BIODETERIOGENS

- **Microbes**
  - Bacteria
  - Cyanobacteria
  - Actinomycetes
  - Funi
- **Plants**
  - Algae
  - Bryophytes
  - Pteridhphytes
  - Spermatophytes
    - a. Gymnosperms
    - b. Angiosperms
      - I. Monocot
      - II. Dicot
- **Animals**
  - Insects
  - Aves
  - Mammals
    - a. Rodents
    - b. Cheroptetra
    - c. Human beings

Deterioration of paper :

Paper is a very complex hygroscopic system with its homogenous network of pores and capillaries of different shapes and diameter. The penetration of water in paper may be inter or intrafibrillar. On undamaged paper, the capillaries are upto 70% of its volume. In deteriorated paper they are filled with fungal mycelium. Fungi possess an array of enzyme systems affecting all book materials. Any paper component can be biologically disintegrated. The acidic pH level is favourable factor for its deterioration. Some fungi are capable of first disintegrating the cellulose of paper fibres resulting into loss of weight. Most fungi settle in the neighbourhood and live at the expense of minor contaminations and organic adhesives. The first metabolites appearing due to disintegration of the cellulose or adhesives are readily assimilated by these fungi. Using them they grow and form their own metabolites affecting paper as chemical reagents or serving as nutrient for the fungi.

Book moulds are for their most part adopted to the bindings, later on other species appear and the penetration of the mould into depth of the material begins. The book mycoflora require greater humidity than the first occupier. The complete degradation of cellulose in paper is affected by two groups of cellulases: endoglucanases which hydrolyze internal β-1,4 glucosidic linkages and exoglucanases which split off mono and disaccharide units from the non reducing end of a cellulose chain.

(a) Discolouration of Paper:

Fungi always bring about changes in paper colour due to their sporulation, biochemical transformations and productions of various fungal metabolites. The species of *Penicillium* and *Aspergillus* constitute 85% of the fungi producing colouring compounds. Many compounds of red shades are produced by *P. spinulosums, P. fellutanums* and *A. fumigatus*. Blood-red, orange-pink and yellow colouring substances are also widely produced by species of the genus *Fusarium* (Nyuksha 1994).

(b) Foxing :

A characteristics brown or rusty spotted discolouration found commonly on old books, manuscripts etc. and even not so old books is referred to as foxing. Various studies have been conducted to tackle this problem like, X-ray spectral analysis, T.L.C. analysis, UV light observation etc. and it has been found that –

1. The presence of iron or similar metals either as concentrations of salts or as an actual speck or particle present in the paper.
2. Fungal infection is the main cause of foxing.
3. The role of fungal activity and iron in the development of foxing is not very clear yet (Dhawan, 1986).
4. Foxing is never, or hardly ever, found on paper which has a high content of mechanical pulp.
5. Paper damaged by foxing does not lose its firmness.
6. Foxing is not found with discolouration of microbial origin.

Deterioration of wood

(a) Mould growth :

Mould growth can be fragmentary, observed on surfaces of damp wood, plaster, wall paper or paint. Moulds like *Cladosporium, Penicillium, Aspergillus, Alternaria* and *Aureobasidium pullulans* commonly occur on surfaces as a superficial growth, causing patchy surface discolourations, usually green, grey or black, occasionally pink or yellow. Some moulds colonize wood whose moisture content is above 20% but appreciable development requires moisture contents at or above the fibre saturation point (28 to 32%). Mould growth is best at temperature ranging from 24 to 30°C. Moulds cause some loss in toughness. They increase the porosity of wood.

(b) Dry rot :

The dry rot fungus *Serpula lacrymans* (syn. *Merulius lacrymans*) mostly attacks soft wood and often causes extensive damage. The term dry rot is misleading as moist conditions are required to initiate the growth and optimum growth is continued at about 20%. Active growth is indicated by silky white sheets or cotton wool-like white cushions with patches of lemon yellow or lilac tinges. White mycelial strands may develop sporophores or fruiting bodies when environmental conditions are unfavourable or nutrients are exhausted.

The dry rot is also called brown rot, a term relating to the manner in which it destroys the cellulose but leaves the lignin largely unaltered so that the wood acquires a distinctive brown colour and the structural strength is almost entirely lost.

(c) Wet rot :

Wet rot of wood is caused by a number of Basidiomycetous fungi. The most important ones are *Coniophora puteana* (Cereebella), Poria fungi (*Poria plaenta*), *Phellinus contiguus, Donkiporia, Pleurotus ostreatus* etc.

Wet rot is also called white rot as it destroys both cellulose and lignin, leaving the colour of the wood largely unaltered but producing a soft felty or spongy texture without cross crack. Many common wet rots may be brown rots, for example *Coniophora puteana* and *Dacrymyces stillatus*. In brown rot wood becomes darker in colour and cracks along and across the grain; when dry such decayed wood will crumble to dust.

(d) Soft rot :

Soft rot is a form of deterioration resembling brown rot which results in the unusual softening of wood. Soft rot is as prevalent as decay; however, it is less damaging and less detectable. Soft rot can be regarded as superficial form of wet rot. It is more usually found in timber having ground contact. More than 100 species are known to cause soft rot. Most of them could cause substantial degradation of sapwood. The most destructive of the species is *Chaetomium globosum*. The fungus causes extensive loss in weight, removing primarily polysaccharides. Cellulose is removed faster than the hemi-cellulose. Soft rot fungi definitely deplete lignin, it is removed much more slowly than the carbohydrates.

Biodeterioration is a formidable problem in the preservation of cultural property in tropical country like India. Nair (1996) discussing the need for further research in the field of biocides to prevent and control the biodeteriogens has stated that vast gaps exist in our understanding on different pests. Effective lethal doses are not known, effect of chemicals on museum objects is not fully understood, the retention effect is not known and use of harmful chemicals and their effect on environment and human health is yet to be worked out.

Nair (1977, 1996) has done pioneering work in the field of biodeterioration particularly in insects and fungi. Agrawal (1984, 1989) and Dhawan (1983-84) worked in the field of biodeterioration. They are associated with prestigious National Research Laboratory for Conservation of cultural property (NRLC), at Lucknow and Indian Conservation Institute (ICI), INTACH Lucknow. Jeyaraj (1991), Thakre and Bhajbhuje (1991), Sethuramalingam *et al.* (1991) are some other scientists associated with fungal studies on museum objects.

1.4 Types of Biodeterioration

(a) Mechanical biodeterioration

In this process material is damaged by physical factors exerted by organisms, for example, disruption of floorboards by dry rot and wet rot fruiting bodies, physical damage caused by rodents, insects, higher plants and birds.

(b) Chemical biodeterioration

In this process, the material is attacked chemically by organisms e.g. oxidation, reduction caused by production of organic acids.

When organisms utilize the material as a food source for its own metabolism, it is called assimilatory deterioration. For example, the breakdown of cellulose in building timber, cotton textiles and rayon; also the moulding of building exterior and interior paints, books, wallpapers. This type of deterioration is also seen in shoes and mattresses.

In dissimilatory deterioration the material is damaged by excretion products liberated by the organisms, not directly concerned with the uptake of nutrients, for example, production of secondary metabolites, toxins, pigments, corrosive waste products, damage to stones by lichens.

Biodeterioration may cause production of bad odour or fouling in certain cases.

Enzymes in Biodeterioration

Wood has been described as an ideal carbon energy source; which suffers from the disadvantage that it is highly polymeric and thus highly insoluble. All wood deteriorating fungi use extra cellular enzymes which are too large to penetrate into the native cell wall. The moisture in the cell lumen swells the cellulose thus rendering the internal surface more accessible for the extracellular enzymes to diffuse into the cell wall. These enzymes catalyze the depolymerization of the different cell wall constituents into water soluble and smaller molecular fragments which diffuse back to the fungus where they are assimilated by the fungus. Enzyme production is greatest in brown rots and least in soft rots, where cellulase production is very low. In brown rot fungi, the hydrolytic enzymes decompose the polysaccharides of the secondary wall. Among the ectoenzymes, cellulase is most important which acts on the polysaccharide.

1.5 Biodeterioration of Monuments

Problems of deterioration may not be only in objects displayed in a museum, it may be present in other cultural properties like monuments, murals and cave paintings.

Different kinds of organisms grow by utilizing the mineral components of stone and its superficial deposits. Their metabolic activity with the excretion of enzymes, inorganic and organic acids and of complex forming substances help in dissolution of the insoluble components of the substratum. Moreover, the growth and the swelling of some vegetative structures like roots and thalli induce physical stresses and mechanical disruptions. In general the kind of biological agent that develops on a stone is affected by the rock type (e.g. alkaline lime stone or acidic sand stone) and its surface conditions, together with the environmental location and climate to which the monument and stone are exposed. The succession of organisms on a monument may be:

a) **Strictly mineral:** The pioneering organisms that develop on a surface which is lacking in organic matter, are the lithophilous photoautotrophs.

b) **Oligotrophic:** A transformed nutrient rich surface with very small amount of organic matter and growth factors, indifferent of their origin either from airborne particles or derived from previously developing organisms, can support the growth of oligotrophic heterotrophs like bacteria, actinomycetes and fungi.

c) **Eutrophic:** The growth and development of successive generations enrich the stone with organic matter and growth factors, the physical surface conditions deteriorate and soiling effect is observed. This provides conditions for growth of microorganisms and spores of Cryptogams. Later on higher plants and arthropods communities may develop. This micro fauna feeds on vegetation and helps in propagation of these plants.

1.6 Fungi affecting Ajanta and Ellora caves:

The earliest mural paintings of the historical period in India are in the Jogimara caves in Madhya Pradesh and are datable back to 2nd century B.C. closely followed by the paintings of Ajanta, which range from 2nd century B.C. to 6th century A.D. The ground used was mud plaster in all these early paintings. At Ajanta, a rough plaster of ferruginous earth admixed with clay and reinforced with vegetable fibres and paddy husk was first laid on the rock surface. Over this was laid a fine thin plaster of clay. A coat of lime was applied over the fine plaster. The pigments used were yellow ochre, red ochre, carbon, lime and gypsum, terreverte and lapis lazuli. The painting was done in a binding medium of animal glue. The use of lime plaster first came into practice around 7th century A.D. and was noticed for the first time at Ellora. Subbaraman (1993) reported growth of micro-organisms like fungi when relative humidity exceeds 65%. On surfaces that are persistently damp, even mosses, lichens and algae can grow. When the plaster contains organic matter such as vegetable fibres and paddy husk, as in Ajanta, can be attacked by insects such as silver fish, beetles etc. and it may be a serious problem.

Paintings at Ajanta and Ellora caves are technically not true **frescoes** but they are **tempera paintings**, viz., paintings done on dry plaster by the application of pigments mixed with gum and glue (Lal, 1966).

According to Lal (1972) Ajanta paintings have been executed on a dry ground of mud plaster. "Flaking of paint in frescoes is not common unless severe conditions exists. The Ajanta paintings, not being frescoes and being executed on dry plaster are much more vulnerable in this respect" (Agrawal, 1975).

Generally it is believed that fungi can cause damage to organic materials but it is not always so. Tilak *et al.* (1970) and Tilak (1991) reported growth of fungus on the paintings of Ajanta and Ellora. He has also studied the aerospora of these caves.

Dhawan *et al.* (1991) reported that the presence of microorganisms causes flaking of paint, powder formation, brown, black and white spots, as well as patches & cracks. They also reported fungi such as *Alternaria, Aspergillus, Cladosporium, Curvularia, Drechslera, Fusarium, Paecilomyces,* and *Trichoderma* causing damage to Ajanta wall paintings.

1.7 Fungi as Biodeteriogens in Museum:

A study on biodeterioration of museum materials by fungi hardly needs any justification. The threat by fungi to cultural property is one of the most wide-spread causes of deterioration of museum objects in tropical climate.

Fungi are achlorophyllous plants and they are unable to synthesize their own food. They are widely distributed throughout the world. Wherever, organic materials are present, the problem of biodeterioration due to fungi takes place. They are usually aerobic i.e. live only in presence of air, while few forms are facultative anaerobes which exist in soil, rocks, air, dead plants or animals. They are also found growing on soured food-stuff, such as breads, pickles, jams and fruits etc. A large number of fungi are parasitic on plant materials. They also cause deterioration of museum objects like paintings, paper manuscripts, wooden objects, leather materials, textiles etc.

Almost all fungi have two phases of their life cycles: vegetative and reproductive. The body of a vegetative fungus is made up of a system of fine filaments called **mycelium**. Each individual of these filaments is called a **hypha** which originates from spores. The hypha may be branched or unbranched. The mycelium spreads its root like threads on host material and produces spore bearing structures called sporangium. Spores may be of different shapes and sizes. They are present in an environment, whenever they find suitable environmental conditions they start germination. Most fungi do not require light for their growth, however, it is needed for a few species. The optimum temperature for their growth varies from species to species but rapid growth is observed at the temperature of 25°C to 35°C (Nair, 1972). However, some fungi can grow at 0° to 60°C. According to Pathak and Dhawan (1989) fungi can be classified into three groups on the basis of limits of the temperature within which they grow. These are—

a. Thermophiles

b. Psychrophiles and

c. Mesophiles

Fungi generally require acidic medium to grow with pH 5.0 to 6.5, although few fungi may grow at much below pH 3.0 and few others even above pH 9.0. Oxygen supply is necessary for their growth.

Humidity is another essential requirement for the growth of fungi. Bilgrami and Verma (1978) reported that 95 to 100 % RH supports best growth of most fungi and humidity below 80 to 85 % is inhibitory for them. However, Pathak and Dhawan (1989) stated that the most suitable RH for the growth of fungi is 75 to 80%. Some fungi, associated with stored products, are able to grow at significantly low RH between 86 to 75%, they are called xerophiles. RH affects the museum materials in different ways. Dryness and wetness of the climate have strong bearing on the life of the museum materials. Therefore, control of climate becomes a very important factor for preserving the museum objects for a long time.

1.8 Fungi under study:

Since fungi inhabit varied objects, isolation of fungi was undertaken from different museum objects. Repeated isolations were made from wooden plaque, musical instruments (Flute, Pavri and Drum etc.), paper manuscripts, miniature paintings, wooden Chinese screen, photographs, leather, herbarium sheets etc. Indoor aeromycoflora of a few institutions have been studied too. Wood and paper materials existing in a museum collection in various forms, are more susceptible to fungi. It was interesting to note that the fungal species present in aerospora of the museum, attacked the objects in the gallery.

The following organisms were isolated from wood and paper materials and were selected for detailed cultural and control studies. These organisms are new reports on mentioned museum object.

Sl. No.	*Name of the organisms*	*IMI No*	*Source*	*Reference*
1.	*Curvularia* state of *Cochliobolus verruculous* (Tsuda and Ueyama) Sivan.	352915	Flute (Musical instrument, Wooden material)	Studies in Museology **26**, 35-36 1992-93 M.S. Univ., Baroda
2.	*Penicillium citrinum* Thom.	352916	Pavri (Musical instrument, Wooden material)	–
3.	*Aspergillus ustus* (Bainier) Thom. & Church.	352924	Wooden Chinese screen	Proceedings of 80th Ind. Sci. Cong., Goa 1993.
4.	*Chaetomium cymbiforme* Lodha	362265	Parsi manuscript (book)	–

C. verruculosus:

It is a member of Fungi imperfecti. It is a phytopathogen and reported from *Typha latifolia, Euphorbia thymifolia,* seed and leaf spot of some pasture plants, sorghum and wheat (Bilgrami *et al.* 1979 and 1981). It is a biodeteriogen and common biopollutant. It is found growing as a saprophytic organism on cellulosic materials like paper, wood, textile etc.

P. citrinum:

It is a member of Ascomycotina. It is reported from grape berries (Tandon *et al.*, 1975), soil (Rao, 1965), betelvine (Singh and Chand, 1973) and jute (Basu and Ghosh, 1950). It is also found growing frequently on dead organic materials.

A. ustus:

It is a member of Ascomycotina. It is reported from soil, cotton seed, pigeon pea, cellulose and compost in shishum forest and *Arachis hypogaea* (Bilgrami *et al.*, 1979). It is also reported from parchment and leather.

C. cymbiforme:

It is a member of Ascomycotina. It is a cellulose degrading fungus generally found growing on cellulosic materials. It is reported from cow-dung (Lodha, 1964).

Based on literature survey *P. citrinum* and *A. ustus* have been reported from some of the museum objects but rest of the two organisms are new reports.

1.9 Cultural and control studies:

Cultural and biochemical studies help us in understanding the nutritional behaviour of the microorganisms and detailed information achieved through these investigations pave way to suggest some suitable control measures.

The studies were conducted on *Cochliobolus verruculosus, Penicillium citrinum, Aspergillus ustus* and *Chaetomium cymbiforme* which caused deterioration of paper and wooden materials. Since no cultural and control studies have been done on these fungal species it was thought desirable to study these aspects in detail

Significant findings:

Of the four organisms selected for detailed studies, only two - *P. citrinum* and *A. ustus* have been reported from some of the museum objects. However, their occurrence on these objects is the first report. No biochemical studies have been reported on these biodeteriogens except *P. citrinum*. Isolation from different museum objects resulted into several new fungi besides these four biodeteriogens. Their presence has also been observed during aeromycological studies. The text incorporates a detailed account of physiological behaviour of four fungi. Effect of different sources has been studied and daily uptake of carbohydrates has been recorded. Most of the studies in this area are limited to reporting of new fungi or other biodeteriogens on different objects. The different chemicals, fungicides and natural plant products have been tried. The effect of these fungi on different woods and papers have been recorded and percentage loss is assessed.

Aerobiology is an emerging branch of Science. Different studies carried out in the field can surely help to preserve the cultural property at one end and human health at another. The fungal flora of different museums and library have been investigated and a direct correlation of some of the biodeteriogens with certain museum objects has been established.

It is true that an artist and a research scientist have much in common. The scientist introduces an idea and the artist an image, both contributions may be unique and new to the world. In general the value of a "good" art also increases with time and serves to generate new art forms and perspectives (Kuc, 1995). The knowledge gained through these experiments will surely help to preserve the cultural heritage in a better way and this will give a glimpse of ancient past to the generations yet to come.

Chapter—2

Biodeteriogens

2.1 Different biodeteriogens

The monuments, sculptures and works of art, exposed to open air have deteriorated, through the centuries due to natural causes. Sun, frost, wind, rain etc. contribute to the gradual process of weathering in which biological activity plays a key role because of its interactions with physico chemical mechanisms and is a main cause of long term deterioration. Biodeterioration is widespread and very familiar process which affects most organic materials and even some inorganic ones. Weathering of building stones is well known since antiquity (e.g. Herodotus reported the deterioration of the Pyramids 2500 years ago). These processes are due to environmental factors. However, industrial and urban activities modify the composition of the atmoshere, resulting in a more aggressive environment. In fact pollution created by numerous industries have accelerated stone weathering since the beginning of this century. Thousands of chemicals harmful to buildings are emitted directly or indirectly into the atmosphere because of man's activities (Saiz-Jimenez 1990).

Deterioration of Taj Mahal, one of the seven wonders of the world has been cause of concern to the conservationists. Several scientists/agenices have studied the problems and advised the government for its conservation. Agrawal (1995) Director of NRLC, Lucknow mentioned that setting up of oil Refinery at Mathura was thought to be a major pollution source, even scienstist described that Taj is suffering from 'stone cancer'. He reported general yellowing of the marble was due to deposit of dirt or soil or due to the alteration of resin polymethyl methacrylate or polyvinyl acetate. Blackish patches were due to growth of algae and small black spots due to dense deposits of mosquito excreta particularly on the north side towards Yamuna river.

The biodeterioration of materials is one of mankind's oldest problems and can be traced back to biblical times or probably earlier. **Biodeterioration** can be defined as any undesirable change in the properties of material of economic importance brought about by the activities of living organisms (Hueck 1968). The concept of biodeterioration is different from **biodegradation** which can be considered as the harnessing by man or natural decay, capabilities of organisms to convert a waste material into more acceptable and manageable form or to produce a useful end product from waste material.

According to geological time scale, life originated 4.6 billion years ago and evidences are available that microorganisms came into existence 3.5 billion years ago in Archaean Era. Life thriving at that time in an ambience of extreme temperature under anoxic conditions was quite different from thermolabile life forms of today. Woese *et al.* (1990) proposed that living beings on this planet should be classified into three domains:

- *Archaea* (Greek, primitive; previously archaeobacteria)
- *Bacteria* (Eubacteria) and
- *Eucarya* (Eucaryotes)

They also traced back the nature of last common ancestor of biokingdom and termed this as 'Progenote' (Prasad & Sarkar 1997).

Biodeteriogens may be bacteria, fungi, algae, lichens, bryophytes, mosses and even higher plants. Termites, insects and beetles may cause decay of wood in buildings and wooden articles. Next to human beings, rodents are the most successful mammal dwellers in buildings. The rats (*Rattus rattus*) and mice (*Mus musculus*) are serious pests in buildings. They cause physical damage owing to their constant gnawing habit. The problems caused by birds in buildings are the fouling of building exterior with dropping etc. Hazira tomb in Vadodara shows damage to the interior walls and ceiling by a mammal bat and pigeons (*Columba livia*).

Various pollutants from the atmosphere are ultimately transferred to buildings and outdoor-exposed statues and monuments, which act as repositories of airborne organic chemicals, accumulating at the surface in zones frequently soaked by rainwater but not washed out. These compounds may influence to some extent the colonization and growth pattern of microorganisms. They can be found both on and beneath surfaces. Microorganisms that inhabit rock substrates are called Lithobionts. They are epiliths, if they live on the surface of the rock, or endoliths, if they live inside it (Golubic *et al.* 1981). The endolithic microorganisms can be divided into : Chasmo endoliths (living in fissures and cracks in the rock), cryptoendoliths (in structural cavities within porous rocks) and euendoliths (penetrating actively into the interior of the rock, forming tunnels). Epilithic as well as endolithic cyanobacteria and algae have been described in nature.

With the advances in the knowledge of cell structure it was realised that all living organisms on earth were composed of one or the other of two types of cells (a) **Eukaryotic** (Gr. eu = true karyon = nucleus) and (b) **Prokaryotic** (Gr. Pro= primitive). The prokaryotic cells has a naked non membrane bound nuclear DNA, and a semi-rigid cell wall of unique composition. The cell size ranges from 1 to 10 mm.

The profound differences between prokaryotes and eukaryotes suggest that a multiplicity of life forms evolved simultaneously in several directions (i.e. polyphyletic evolution) from a variety of primitive forms.

American ecologist Robert Whittaker (1969) proposed a 5 Kingdom system for living beings. These are Monera, Protista, Plantae, Fungi and Animalia. This system of classification is based on level of cellular organisations and mode of nutrition. In addition to plants and animals a German biologist Haeckel (1866), proposed a third Kingdom 'Protista' (Protisto=very first) to include oganisms which lacked tissues viz. algae & protozoa etc. The protists themselves were later divided into lower and hjigher protista. The lower protista which included prokaryotes-bacteria and blue green algae was named the fourth kingdom '*Monera*' (Syn = Mychota) (Copeland 1956, Barkley 1968). Viruses, Viroids and recently discovered proteinaceous infective particle commonly known as slow viruses - Prions being non cellular are excluded from this scheme of classification. Living organisms can be divided into two super kingdoms, the *Prokaryonta* and *Eukaryonta*. It is proposed that eukaryotes evolved by the symbiotic association (living together, one inside the other) of two or more kinds of prokaryotes.

Cellular organisms are classified as follows :

I. Prokaryonta (without true nucleus)

1. **Archaebacteria** - live in anaerobic environments, produce methane by reduction of CO_2 can live in extreme saline conditions, hot and acidic environments.
2. **Eubacteria** - True bacteria discovered by Van Leeuwenhoek in 1676. (Cyanobacteria (blue green algae), Spirochaetes Rickettesiae, Phytoplasmas, Spiroplasmas and Actinomycetes are also members of Prokaryonta).

II. Eukaryonta (cells with true nucleus)

3. **Protista** - This group includes unicellular algae, protozoans and slime molds.
4. **Mycophyta** - Includes true fungi (non chlorophyllous thallophytes)
5. **Plantae** - includes green, brown and red algae, bryophytes, pteridophytes and spermatophytes)

6. **Animalia** - includes all multicellular animals.

2.2 Fungi as biodeteriogens

Man's interest in fungi started with the observation of the beautiful umbrella-shaped mushrooms and toadstools growing on soils forming 'fairy rings'. Fungi have been objects of wonder and speculation eversince their role in biodeterioration is well known. Etymologically, mycology means the study of mushrooms (Gr. mykes= mushroom, logos = discourse).

Fungi are classified into division *Mycota* which has two sub-divisions *Myxo mycotina* and *Eumycotina*. More than 100,000 species of fungi are now included in a separate kingdom, Kingdom **Myceteae** (The Fungi). Alexopoulos (1962) defined the term fungus (Pl. Fungi) to the nucleated spore-bearing, achlorophyllous organisms which generally reproduce sexually and asexually, and whose usually filamentous, branched somatic structures are typically surrounded by cell walls containing cellulose or chitin, or both.

True fungi are heterotrophs, either parasites or saprophytes. Obligate saprophytes or saprobes live only on dead animal or plant tissues and are incapable of invading living tissues. These can be artificially cultured. Fungi produce extracellular enzymes convert the insoluble food into soluble form which are then absorbed (Dube 1990).

The Italian mycologist P.A. Saccardo collected and brought together the scattered knowledge on systematic mycology under his massive 25 volume work, the *Sylloge Fungorum Omnium hucusque cognitorum* (1882-1931), whose first volume appeared in 1882 and the 25th in 1931. It was a Herculean job! The true fungi were divided by Saccardo into four classes on the basis of presence and absence of septa and characteristic, sexually produced spores.

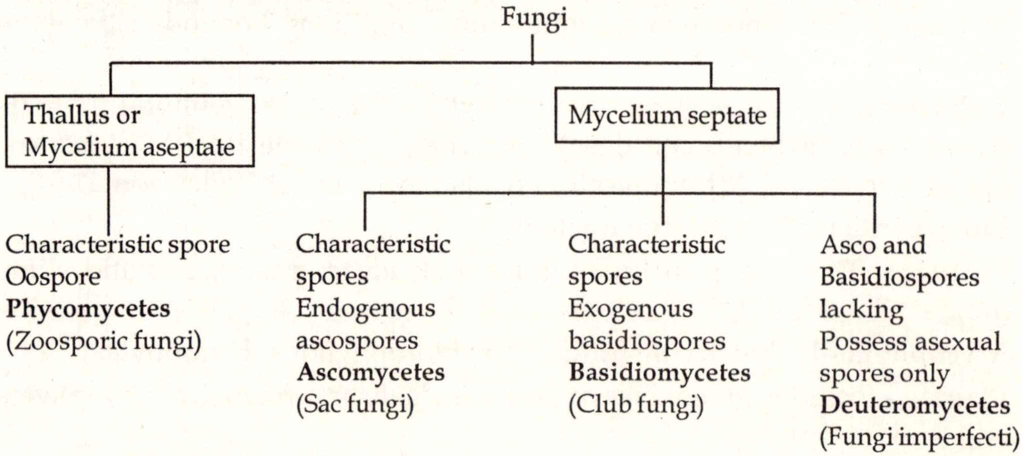

The Phycomycetes or primitive algal fungi, are sometimes called as 'lower fungi' and the fungi belonging to Asco-Basidio- and Deuteromycetes as the 'higher fungi'.

Class Phycomycetes has two sub-classes Oomycetes and Zygomycetes. Sparrow (1958) recognised four 'galaxies' in aquatic Phycomycetes and named them as four classes. Alexopoulos (1962) used this nine class classification in his book 'Introductory Mycology'. The nine classes are 1. Chytridiomycetes, 2. Hyphochytridiomycetes 3. Plasmodiophoromycetes, 4. Oomycetes 5. Zygomycetes, 6. Trichomycetes, 7. Ascomycetes, 8. Basidiomycetes and 9. Deuteromycetes (Fungi imperfecti).

The next important change came in 1966 when G.C. Ainsworth, Director of Commonwealth Mycological Institute (CMI now known as IMI) proposed a classification of Fungi as follows :

Kingdom - Mycota (Fungal Kingdom)
(Classification as proposed by Ainsworth 1966)

Division	*Division*
Myxomycota (4 classes) (Plasmodial forms)	**Eumycota** (5 sub-divisions)
called slime molds	(True fungi, are usually mycelial)
1 - Acrasiomycetes	1 - **Mastigomycotina**
2 - Hydromyxomycetes	- Chytridiomycetes
3 - Myxomycetes	- Hyphochytridiomycetes
4 - Plasmodiophoromycetes	- Oomycetes
	- Plasmodiophoromycetes
	2 - **Zygomycotina**
	- Zygomycetes
	- Trichomycetes
	3 - **Ascomycotina**
	- Hemiascomycetes
	- Loculoascomycetes
	- Plectomycetes
	- Laboulbeniomycetes
	- Pyrenomycetes
	- Discomycetes
	4- **Basidiomycotina**
	- Teliomycetes
	- Hymenomycetes
	- Gasteromycetes
	5- **Deuteromycotina**
	- Blastomycetes
	- Hyphomycetes
	- Coelomycetes

Ainsworth treated fungi as separate kingdom. Myxomycota or slime molds are divided directly into four classes and Eumycota or true fungi first into 5 sub-divisions and then into classes. Because of certain common characters plasmodiophoromycetes finds place in both divisions. The Deuteromycotina, though given a rank equal to Ascomycotina and Basidiomycotina can not be equated with the latter. Fungi imperfecti are supplement to them, both in status and nomenclature. Perfect states of a large number of Deuteromycetous fungi are discovered and they belong to Asco or Basidiomycotina.

1. Mastigomycotina

Commonly known as zoosporic fungi. The zoospores are uniflagellate or biflagellate. The fungi have coenocytic thallus, parasites or living saprophytically, usually they require water/moisture for their growth. Act as a decomposer of plant or debris and their role has been seldom reported as biodeteriogens.

2. Zygomycotina:

Coenocytic fungi which lack motile cells and produce zygospores as perfect or sexual spore. Mycelium extensive, produce sporangia and aplanospores. Common biodeteriogens are *Absidia, Chaetostylum, Circinella, Cunninghamella, Mortierella, Mucor, Rhizopus, Syncephalastrum* and *Thamnidium*.

3. Ascomycotina:

Septate vegetative mycelium, produce large fruit bodies - ascocarp except in Hemiascomycetes. Fungi possess an ascus containing usually 8 ascospores produced endogenously after Karyogamy and meiosis.

Common biodeteriogens are *Agyrium, Ascobolus, Ascotricha, Ceratocystis, Chaetomium, Cochiliobolus, Mycotrichum, Nectria, Nectriella, Peziza, Pleospora, Pyronema, Rhodotorula, Sclerotinia, Sordaria, Thielavia* etc. (Nyuksha 1993).

4. Basidiomycotina:

This is a highly evolved group of fungi. Commonly known as club fungi. Produce basidiospores (usually 4) externally on the basidium. They have highly developed dikaryophase. Presence of dolipore septum in mycelium is yet another feature. These are common wood rotting fungi. Several wooden objects show deterioration and different buildings with wooden beams and doors may be attacked by these fungi. Sometimes fruiting bodies called sporophores or basidiocarps are produced. Common biodeteriogens of this group are *Amyloporia Botryohypochnus, Coniophora, Coprinus, Dacrymyces, Donkiporia, Gloeophyllum, Fibroporia, Hirschioporus, Lentinus, Paxillus, Phellinus, Pleurotus, Serpula, Stereum.*

5. Deuteromycotina:

These fungi are also known as fungi imperfecti. They reproduce by asexual means, usually by conidia. The mycelium is yeast like pseudomycelium in Blastomycetes or true, branched and septate in Hyphomycetes and Coelomycetes.

Mycelium sterile or bearing spores directly or on special branches (sporophores) which may be variously aggregated but not in pycnidia or acervuli - **Hyphomycetes**. These fungi were placed in Mycelia sterilia and Moniliales by Saccardo (1886). If spores are produced in pycnidia (Sphaeropsidales) or in acervuli (Melanconiales) the fungi are placed in **Coelomycetes**. The genera and species are identified by the colour, shape and septation pattern of their conidia. Many biodeteriogens belonging to this group are *Acremonium, Alternaria, Botrytis, Cladosporium, Curvularia, Fusarium, Nigrospora, Phoma, Trichoderma, Stachybotrys, etc.*

2.3 Characters of common fungal biodeteriogens

A. ZYGOMYCOTINA :

1. *Absidia* van Tieghem

The mycelium comprises of stolons and rhizoids but with the difference from *Rhizopus*. Here the sporangiophores arise in groups of 2 to 5 from internodes of the stolons and not opposite the rhizoids. The sporangia are pyriform and thin walled, columella has an apical projection. Circinate appendatges arise from one or both the suspensors enveloping the zygospore which are borne on the stolons.

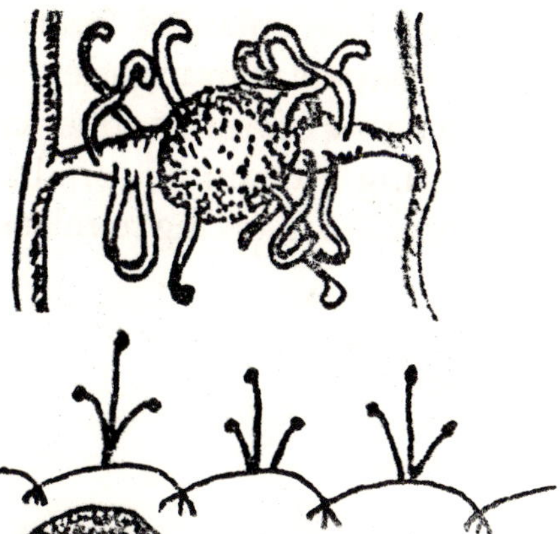

1. *Absidia*

2. *Mucor* Mich. ex Fr.

This is a large genus of family Mucoraceae. The hypae are coarse, coenocytic and richly branched, usually tapering to fine points. Sporangiophore bear typical collumellate globular sporangium. The zygospores are always brown to black, with rough blunt projections or warts.

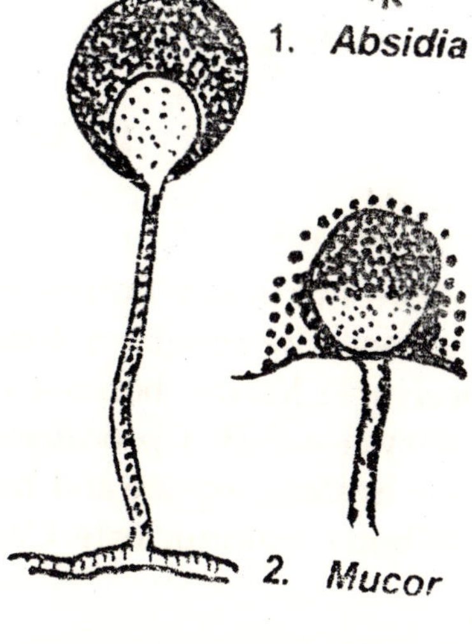

2. *Mucor*

3. *Rhizopus* Ehrenb. ex Corda

This is another member of Muco-rales. Mycelium has root like rhizoidal mycelium, and aerial hyphae called stolons. Sporangiophores are shorter, stouter and stiffer than those of *Mucor*, arise from the Junction of the stolons and the rhizoids. The sporangiophores are dry and easily blown by wind.

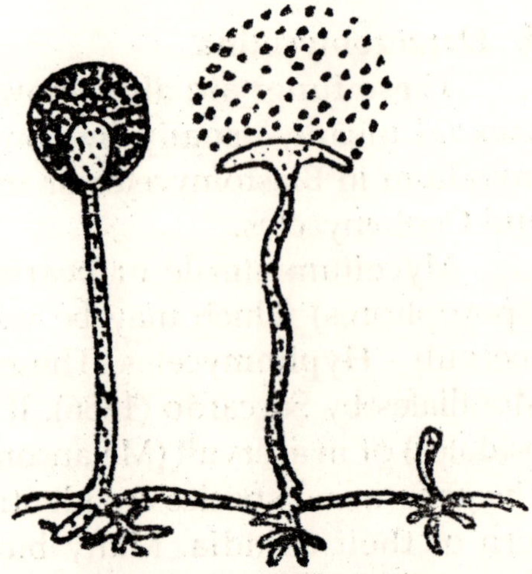

3. Rhizcpus

4. *Syncephalastrum* Schroet.

Mycelium growing rapidly, abun-dantly branched; sporangiophores erect, branched, tips enlarged, bearing a head of rod-shaped sporangioles, each pro-ducing a row of nearly spherical spores, resembling a chain of conidia, wall of sporangiole dissolving to release spores.

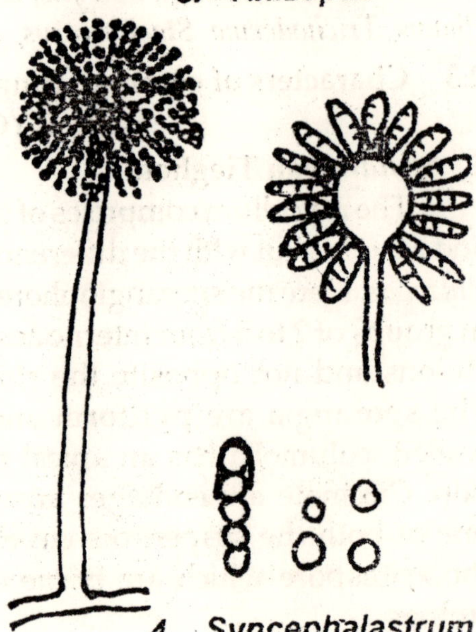

4. Syncephalastrum

B. ASCOMYCOTINA

5. *Aspergillus* Link.

It belongs to family Eurotiaceae of Plectomycetes. The fungus is found in any conceivable habitat because of the bizarre enzymes that it produces. The hyphae are hyaline, septate and branch-ed. The cells are multinucleate. Conidial stage is dominant, sexual stage is known as *Eurotium, Emericella* and *Sartorya.* Conidiophore terminates into a swollen vesicle, on which bottle shaped structures called sterigmata or phialides bear chains of conidia.

6. *Chaetomium* Kunze ex Fr.

Chaetomium is an important cellulose degrading fungus and is found growing on paper, fabrics etc. It belongs to family Melanosporaceae of class Pyrenomycetes. The fungus produces perithecium with long coiled hairs. The asci are stalked, club shaped, cylindrical or oval and produced in a basal tuft. Asci are evanescent and ascospores are released passively. Ascospores are dark and inequilateral.

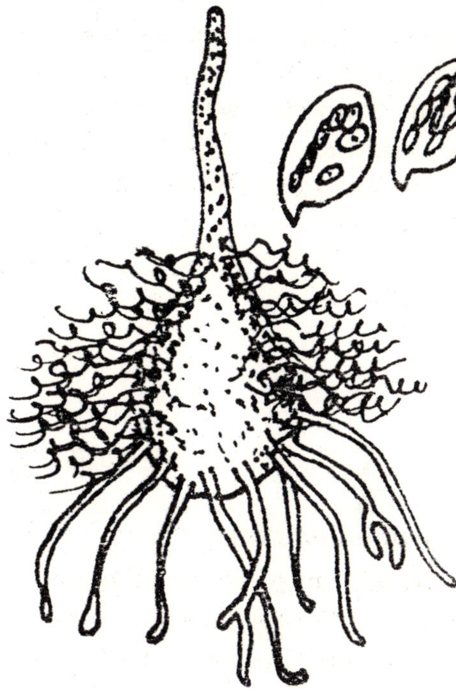

6. *Chaetomium*

7. *Penicillium* Link.

The fungus produces a conidial apparatus which resembles a brush or broom, is called a 'penicillus'. Long septate conidiophore can arise from any cell. Conidiophores branch once or twice and produce sterigmata or rammi and finally phialides. The colour of conidia green or blue. The sexual stage is called *Talaromyces* or *Eupenicillium.* The asci are globose and evanescent.

8. *Sordaria* Ces ex de Not.

It belongs to Sordariaceae family of class Pyrenomycetes. It produces perithecium which is dark and pear shaped. Branched septate mycelium with multinucleate cells. Reproduction is by ascospores. The asci arise from base in a tuft.

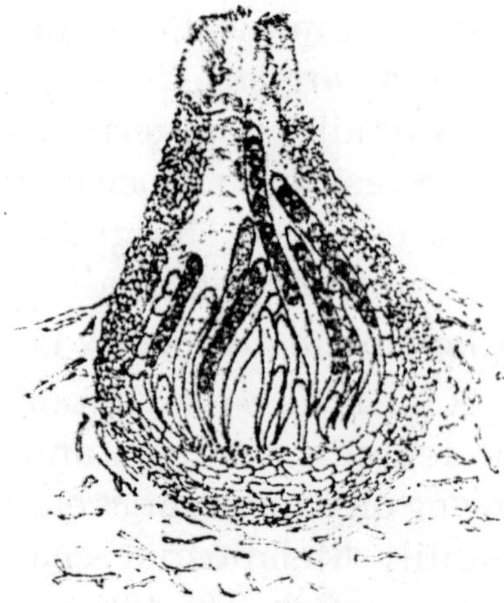

8. Sordaria

C. DEUTEROMYCOTINA

9. *Acremonium* Link.

Mycelium prostrate, slender, producing simple upright conidiophores, conidia hyaline, 1-celled, borne singly, apically.

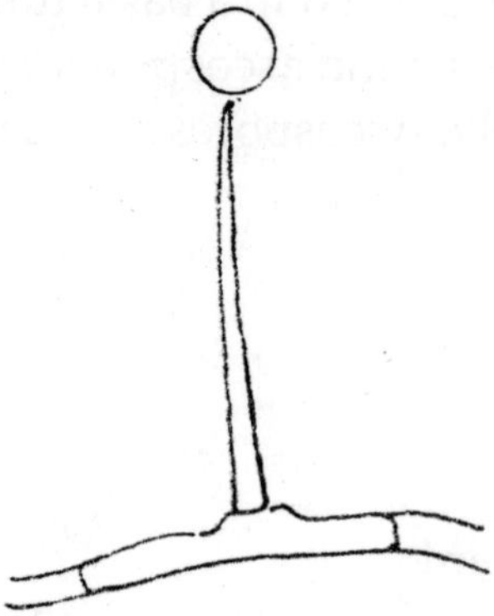

9. Acremonium

10. *Alternaria* Nees

Conidiophores dark, simple, rather short or elongate, typically bearing a simple or branched chain of conidia. Conidia dark typically with both cross and longitudinal septa, muriform, catenate, mostly obclavate and rostrate, frequently borne acropetally in long chains, less often borne singly and having an apical simple or branched appendage.

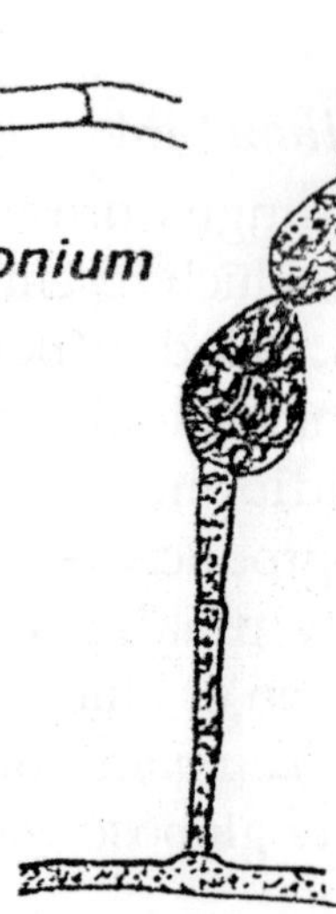

10. Alternaria

11. *Botrytis* Micheli ex Fries

Colonies effuse, often grey powdery under the low-power binocular microscope; stout brown conidiophores are seen supporting glistening heads of pale conidia, mycelium immersed or superficial; conidia solitary, acropleu rogenous, simple, colourless.The conidia look like grape bunch, conidia single celled, occasionally 2 or 3 cells. Sclerotia appear after a few days, which are irregular in shape and serve to perennate the fungus. *Botrytis* is the conidial stage of *Sclerotinia,* a Discomycete (Ascomycotina).

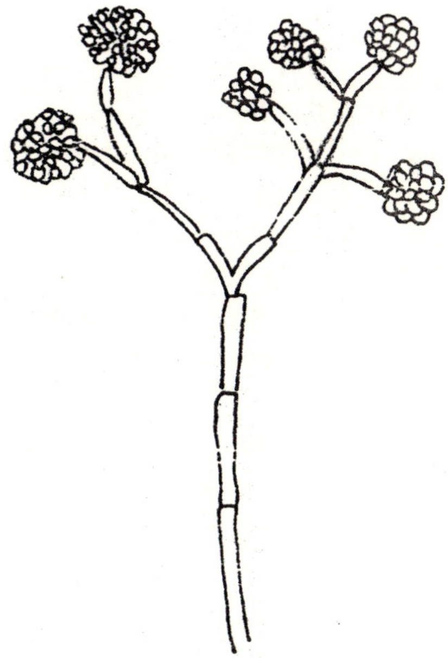

11. Botrytis

12. *Cladosporium* Link ex Fries

Colonies effuse or occasionally punctiform, often olivaceous but also sometimes grey, buff, brown or dark blackish brown, velvety, floccose or hairy. Conidiophores macronematous or semimacronematous or sometimes also micronematous, dark, branched variously near the apex or middle portion, clustered or single; conidia dark, 1 or 2 celled, variable in shape and size, ovoid to cylindrical and irregular, some typically lemon-shaped, parasitic on higher plants or saprophytic.

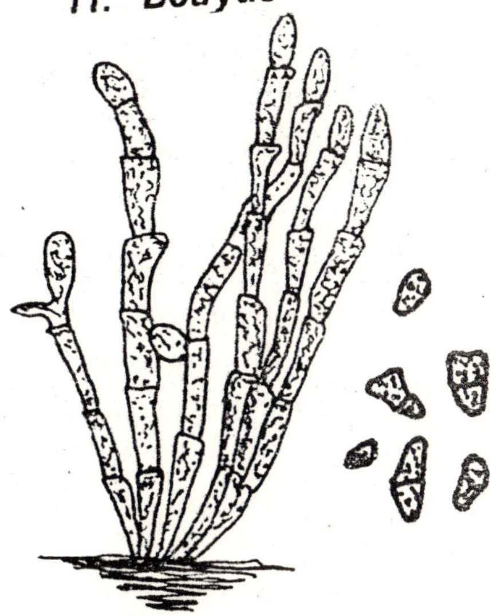

12. Cladosporium

13. *Curvularia* Boedijn

Colonies effuse, brown, grey or black, hairy, cottony or velvety. Mycelium dark, branched, septate, immersed in natural substrata. Conidiophores brown, simple or sometimes branched, conidia dark, and cells lighter, 3-5 celled, more or less fusiform, typically bent or curved, with one or two of the central cells enlarged, smooth or verrucose. The conidiophore forms a sympodula bearing a cluster of conidia, which may be compact or lax. The detached conidium leaves a knob-like scar on the conidiophore on the 'bends' or geniculations, formed due to the characteristic growth as a sympodula. The conidia of *C. lunata*, the type species, is crescent or half moon shaped. Some species are active degraders of cellulose and damage textile.

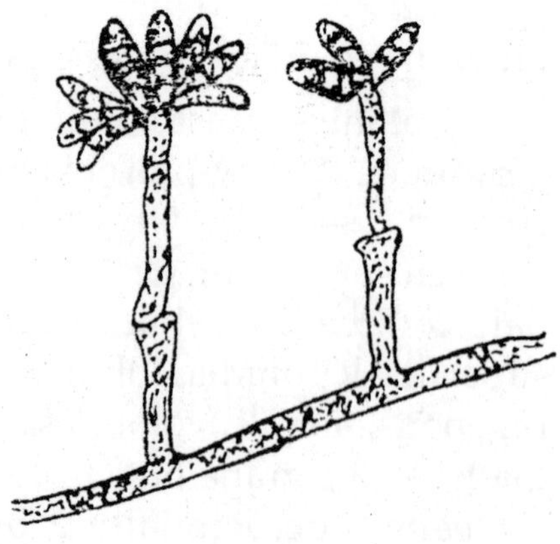

13. *Curvularia*

The perfect stage is *Cochliobolus*.

14. *Drechslera* Ito

Colonies effuse, grey, brown or blackish brown, often hairy, sometimes velvety. Mycelium immersed. Conidiophores macronematous, mononematous, sometimes caespitose, straight or flexuous, often geniculate, unbranched. Conidia solitary, catenate or forming secondary conidiophores which bear conidia, acropleurogenous, simple, straight or curved, clavate, cylindrical rounded at the ends, ellipsoidal, fusiform or obclavate, straw coloured or pale to dark brown, smooth rarely verruculose, pseudo septate.

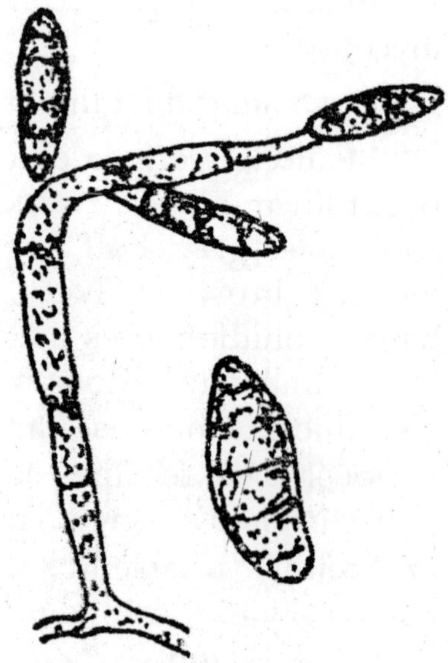

14. *Drechslera*

Its perfect stage is *Pyrenophora*.

15. *Fusarium* Link.

Mycelium extensive and cottony in culture, often with some tinge of pink, purple or yellow, in the mycelium or medium. The hyphae are hyaline, branched and septate. Conidia are formed in slimy, effuse sporodochia called

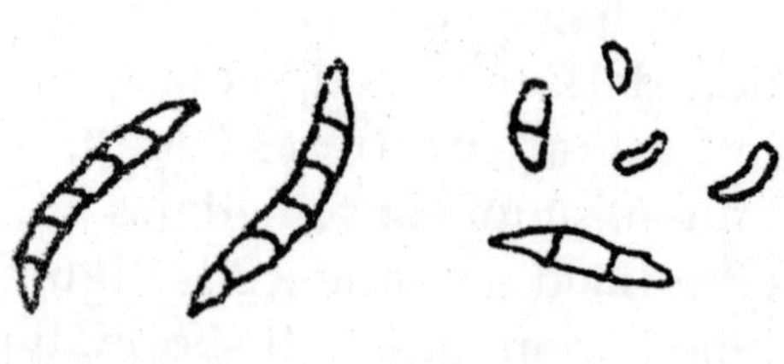

pinnotes. Conidiophores short, simple or branched, septate, bearing a terminal phialide. The phialides are subulate i.e. widest at base, narrowing to a point like an awl. The conidia are of two types **microconidia** (one celled, oval or comma shaped, pyriform or elongate) and **macroconidia** (which are hyaline or pale, spindle shaped with pointed ends; hence the name fusus = spindle), 3 to many septate with a distinct pedicellate base. The macroconidia with a distinct 'foot cell' bearing some kind of heal, is the most distinct feature of *Fusarium*. Chlamydospores are formed on the hyphae or on macroconidia.

This fungus belongs to family Tuberculariceae because of production of sporodochia. The perfect stages belong to several ascomycetous genera viz. *Hypomyces, Gibberella, Nectria* and *Calonectria*.

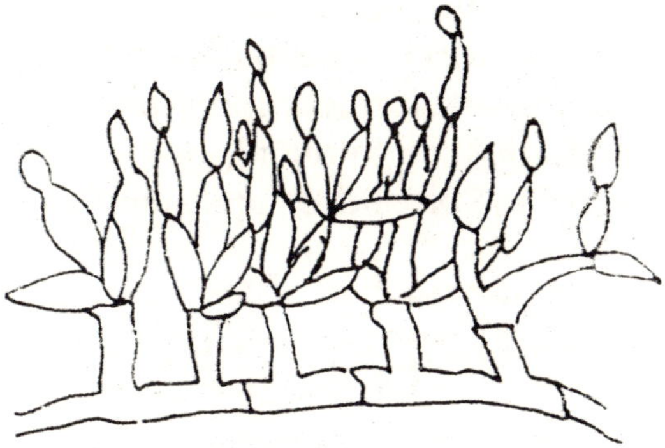

15. *Fusarium*

16. *Nigrospora* Zimmermann

Colonies at first white with small shining black conidia. Conidia solitary with ampulliform or subspherical colourless conidiogenous cells.

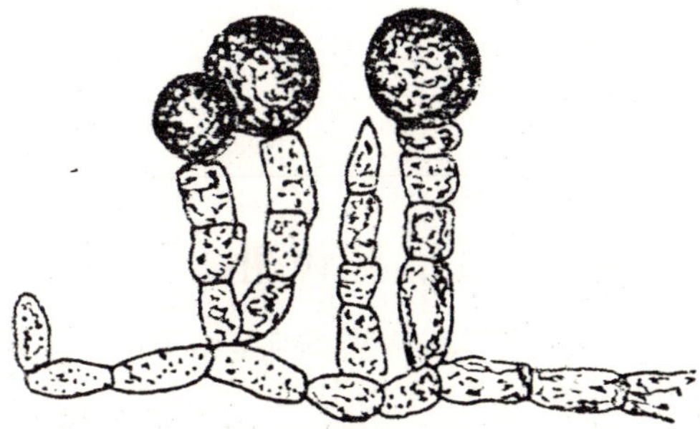

16. *Nigrospora*

17. *Phoma* Desm.

Fungus produces dark fruiting bodies pycnidia, ostiolate, lenticular to globose, immersed, erumpent or with a short beak piercing the epidermis; conidiophores short or obsolate; conidia small, 1-celled hyaline, ovoid to elongate.

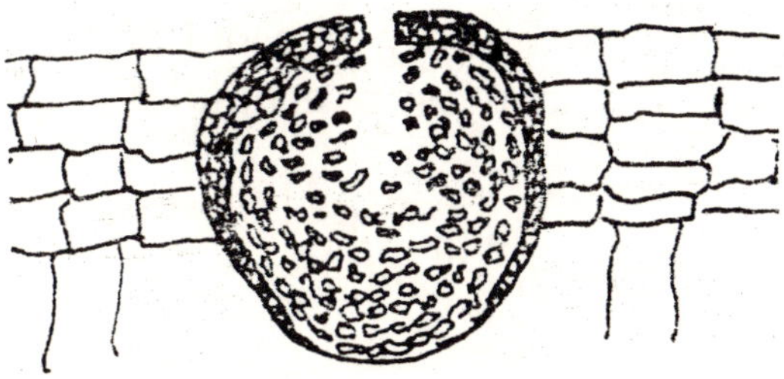

17. *Phoma*

18. ***Stachybotrys* Corda.**

Conidiophores dark, simple, except apex where a cluster of thick, short phialides are formed; conidia dark, 1-celled, globose to ovoid, borne at the apex of the phialides, not catenulate.

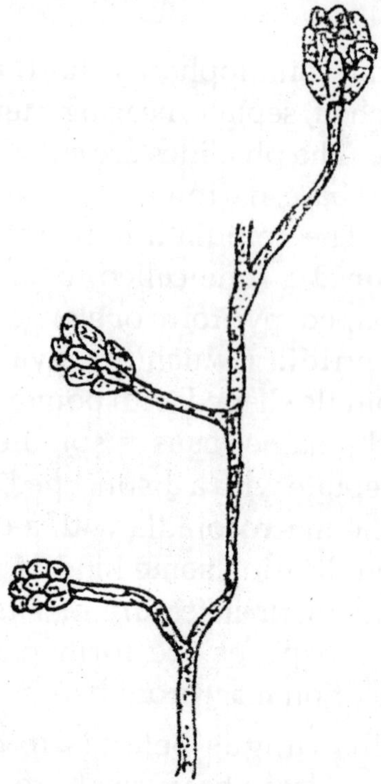

18. Stachybotrys

19. ***Stemphylium* Wallr.**

Conidiophores dark, mostly simple, short to long, bearing a single, terminal conidium or successive conidia on new growing tips, conidiophore proliferating through old conidial scar; conidia dark, with cross and longitudinal septa, variable in shape, frequently globose to broadly ellipsoid or ovoid, smooth, verrucose or echinulate.

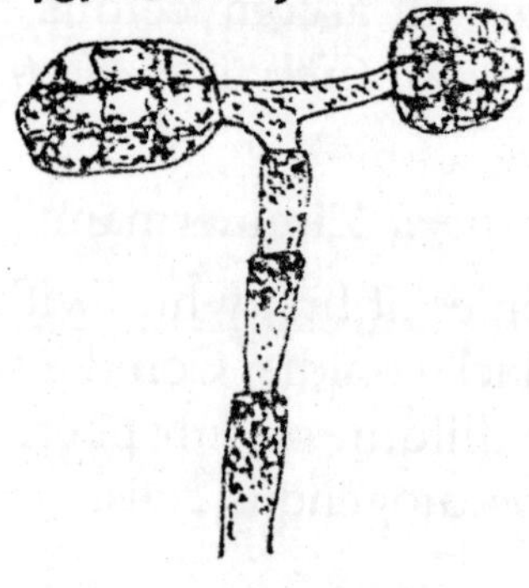

19. Stemphylium

20. *Trichoderma* Pers.

Conidiophore hyaline, upright, much branched, not verticillate, phialides single or in groups; conidia hyaline, 1-celled, ovoid, borne in small terminal clusters; usually recognised by its rapid growth and green patches or cushions of conidia.

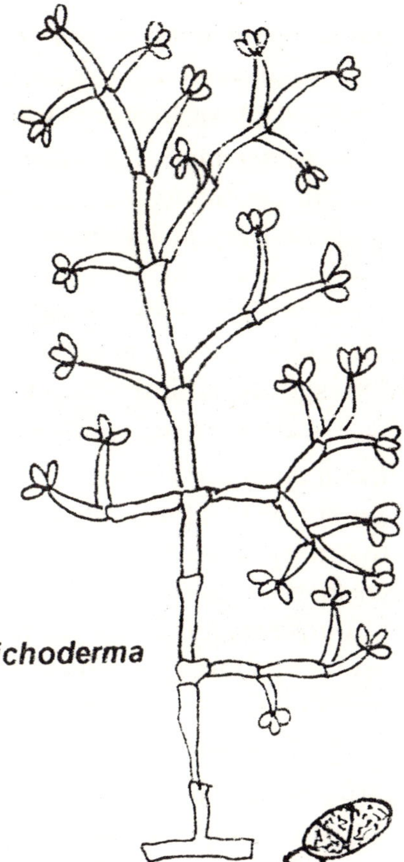

20. *Trichoderma*

21. *Ulocladium* Preuss

Colonies effuse, brown, olivaceous brown, dark blackish brown or black. Mycelium partly superficial, partly immersed. Conidia solitary, broadly ellipsoidal or obvoid, sometimes clavate, pyriform or subspherical but not obclavate or rostrate, frequently with a minute projecting hilum, dark blackish brown, smooth or verrucose, with transverse and usually also longitudinal or oblique septa; septation sometimes cruciate.

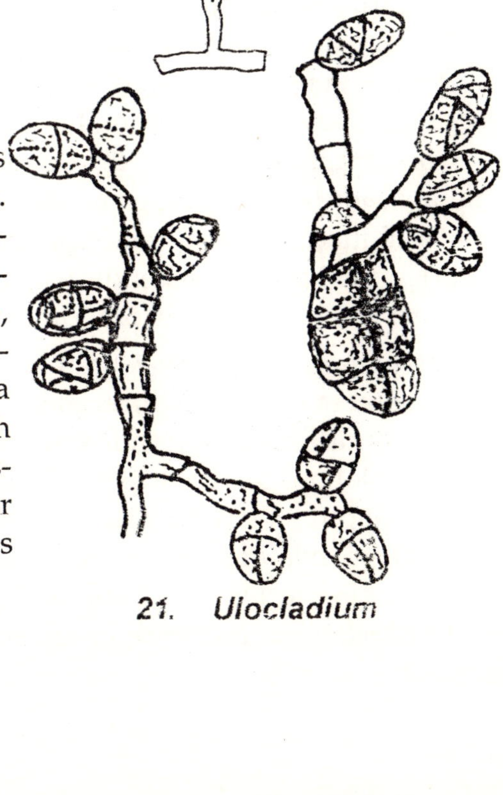

21. *Ulocladium*

2.4 Morphological Characters of Selected fungi

In order to understand the morphological nature of the four organisms, light microscopic studies were conducted (Plate-II)

A. *Cochliobolus verruculosus* (Tsuda & Ueyama) Sivan.

Curvularia verruculosa anamorph, member of Dematiaceous Hyphomycetes, classified under Deuteromycotina/Fungi imperfecti.

Colonies grey or black, velvety, mycelium immersed in agar, stromatic, conidiophores straight or flexuous, often geniculate; conidiogenous cells integrated, terminal or intercalary, sympodial, cylindrical or occasionally swollen; conidia, simple, solitary, curved, 26.6 μm X 7.8 μm in size 3 celled or 4 celled with 3 transverse septa, middle 2 cells dark, wall of spore verrucose.

B. *Penicillium citrinum* Thom.

A member of Plectomycetes in Ascomycotina. Colony in Czapek's agar showing restricted growth. Fruiting surface bluish green, velvety, reverse of light brown shade. Conidiophores erect, comparatively short 120 μm in length, arising as a dense strand from the substratum or the basal felt. Penicilli typically consisting of a terminal cluster of 3 to 5 metulae, each bearing a crowded cluster of parallel sterigmata and chains of conidia. They are small 2.4 to 3.4 μm, globose with walls smooth.

C. *Aspergillus ustus* (Bainier) Thom. & Church.

Member of family Eurotiaceae in class Plectomycetes of Ascomycotina. Perfect state genera is termed as *Emericella*. Colonies on Czapek's agar fast spreading, first white becoming light brown.

Conidiophores produced from aerial hyphae, reverse brown in colour. Conidial heads hemispherical variable in size. Conidiophores 125x2.5 μm. Vesicles 8 to 15 μm in diameter; sterigma colourless, radiate loosely, arranged in two series, primary and secondary sterigmata present; conidia globose 2.6-5.2 μm rough, echinulate. Cliestothecia not seen in culture.

D. *Chaetomium cymbiforme* Lodha.

A member of order Sphaeriales in Pyrenomycetes of Ascomycotina.

Perithecia ostiolate, terminal hairs not dichotomously branched, arcuate. Perithecia globose to avoid 240x290 μm. Lateral hairs light coloured, loosely one to two spiral, convolute at tip; ascospores olive brown, flattened at one side 4 x 2 μm, hyaline when young, asci not observed.

According to Dr. P.F. Cannon, C.M.I., Kew, England this species is related to *Chaetomium trilaterale* group but it does not produce bright coloured pigments.

Chapter—3

Investigation Techniques

Problems of biodeterioration due to fungi are acute and it is difficult to ascertain an individual fungus attacked on museum objects without proper identification. For the identification of fungus isolation and purification of fungal culture is essential. To study biodeterioration it is necessary to adopt scientific approach. Some of the techniques employed are discussed here in detail.

3.1 Preparation of Culture Media:

Some of the common criteria for classifying media are their chemical composition, physical state and their empirical use. In fact, every medium is designed for a definite use and hence its physical and chemical characteristics must conform to its application and function. According to their use, media may be categorised into the following types:

(1) *Routine laboratory media*: These are media with certain complex raw materials of plant or animal origin such as yeast-extract, malt extract, peptone etc., and are employed for routine cultivation and maintenance of a wide variety of fungi.

(2) *Enriched media*: These media are prepared by supplementing the routine laboratory media with some specific substances to meet the nutritional requirements of more fastidious organisms and are employed for their cultivation.

(3) *Selective media*: These media facilitate the isolation of a particular group of organisms or species from a mixed inoculum. Such media contain substances which inhibit all except the desired organism.

(4) *Differential media* : Supplemented with certain reagents or chemicals, these media aid in differentiating between various kinds of organisms on the basis of visible differences in their growth patterns. However, such type of media are used more often in bacteriological laboratories.

(5) *Assay media* : These types of media are specifically employed for the assay of vitamins, amino acids, antibiotics, disinfectants etc., and are of definite composition.

(6) *Biochemical media*: Such media are generally used for the differentiation

of micro-organisms on the basis of their biochemical activities, and are helpful in the study of their metabolic processes.

According to chemical composition, media are classified into the following types:

(a) **Natural media:** A natural medium comprises of entirely complex natural products of unknown composition. The raw materials of a natural medium may be of plant or animal origin, and some of the common ingredients employed for this purpose include extracts of plant and animal tissues, e.g., fruits, vegetables, egg, milk, blood, body fluids, yeast, malt and manure extracts etc. Obviously, the chemical composition and concentration of a natural medium is not well defined. On account of their complex nature, these media are able to support a variety of organisms, and hence are quite useful for routine laboratory cultures of fungi. Brefeld (1881), who was one of the pioneers in the field of fungal culture was so much impressed by the utility of some natural media he used, that he considered atleast one of them, viz. manure extract, of universal applicability for culture of fungi. Other advantages of natural media are their low cost and easier method of preparation. However, these media have certain limitations too. Due to their complex nature, their chemical composition and concentration can not be controlled. This limits their use to routine culture of fungi only, as investigations pertaining to fungal nutrition and metabolism can hardly be carried out on such media.

(b) **Semisynthetic media:** These media are so designed that some of their constituents are of known chemical composition, while others are derived from some natural sources with unknown composition. The chemical make-up of a semisynthetic medium is, thus, only partly known. Consequently, only a limited amount of control may be exercised on the composition and concentration of a semi-synthetic medium, by making necessary changes in the chemically known fraction. Semisynthetic media have also limited application in physiological studies on fungi, and can best serve as a routine medium. Potato-dextrose agar is one of such accepted and popular media. Lilly and Barnett (1951) consider all agar-solidified media as semisynthetic ones, because their exact chemical make-up is partly obscured by the addition of agar-agar.

(c) **Synthetic media:** These are chemically defined media of known composition and concentration, and are exclusively composed of pure chemical substances. However, absolute purity of the ingredients is seldom achieved, although substances of only analytical reagent quality are used for such purposes.

On account of their known composition as well as being in solution, these media are quite useful for nutritional and metabolic studies of fungi. The composition of these media may be amended as per requirement and as such they may be simple or complex in make-up. A simple synthetic medium contains a single carbon and energy source, a nitrogen source, generally as ammonium salt, some sulphur and phosphorus sources and various minerals. All these ingredients are dissolved in a buffered aqueous base. However, for more fastidious organisms, a complex synthetic medium is designed by incorporating some additional factors such as certain vitamins, amino-acids, purines, pyrimidines etc., or by employing a multitude of carbon and nitrogen sources together.

Preparation of a medium is quite an elaborate procedure and involves several steps. Some of the important steps are detailed below:

(A) Dissolution of prescribed amount of ingredients in an appropriate volume of distilled water. In case of a natural medium, this step involves the preparation of a decoction or infusion of the requisite material.

(B) Adjustment of the culture solution to a suitable pH.

(C) Dispensation or distribution of medium into suitable containers.

(D) Sterilization of the medium by an appropriate method.

Adjustment of pH :

Fungi, like other micro-organisms are very much susceptible to the reaction of the surrounding medium upon which their growth and survival depends. Some of the fungi are very much specific with regard to the reaction of the medium while others tolerate a wider range of acidity or alkalinity. However, all have some particular reaction, at which they attain their optimal growth. Fungi exhibit differences over bacteria and actinomycetes in that they are comparatively better adapted to invade acid environment but this does not mean that fungi are restricted to such environments only.

The reaction of any liquid is expressed in terms of hydrogen ions present in it, since that is easier to be measured; similarly for the reason of practical convenience hydrogen ions are expressed on logarithmic scale or pH scale. Thus, pH value of a liquid is the logarithm of the reciprocal of the hydrogen ion concentration, i.e.,

$$pH = \log \frac{1}{H^+} \qquad \text{or } -\log10\,[H^+]$$

$$\text{For neutral water, } pH = \log \frac{1}{10^{-7}} = 7.0$$

(H^+ of pure water at 22°C is 1×10^{-7} g ions per litre). However, for a clear understanding of the pH scale, the following two points must be emphasised:

(i) Since it is a logarithmic scale, a change in one unit of pH will actually represent a tenfold shift in the hydrogen ion concentration. For example, a liquid at pH 4.0 is ten times more acidic than at pH 5.0 or it is ten times more alkaline at pH 10.0 than what it will be at pH 9.0.

(ii) On account of its reciprocal nature, the lower the pH value, the higher will be the acidity. Actually, pH values below 7.0 indicate acidity while beyond 7.0 indicate alkalinity.

For measurement and adjustment of pH of culture-media several methods are employed, viz. (a) electrometric method, (b) indicator dye method, (c) colorimetric method. Among these techniques, the electrometric method is the most accurate in which potentiometers are employed for pH determination. However, use of a wide range of indicator dyes and indicator impregnated pH-papers are now preferred for routine purposes. The colorimetric method, although employed quite frequently due to its simplicity, is the least accurate among these three techniques.

While adjusting the pH of the culture medium at the optimum value for the organism concerned, it has to be kept in view that micro-organisms attack the medial constituents and thus are able to cause a shift in the acidity or alkalinity of the medium, sufficient to even kill them. To avoid such a situation, or to minimize pH changes in the medium, certain buffers are incorporated in the medium. These buffers are compounds, which have the capacity to resist pH changes. Two such compounds, which are widely employed as buffers are K_2HPO_4 and KH_2PO_4. Some other ingredients of the media like peptone also possess some buffering capacity.

3.2 Isolation Techniques:

A scientific method for the isolation of fungi from museum objects is as follows. First the infected object is cleaned with 90% ethanol and when fungus growth reappears, it is isolated in a culture tube containing PDA (Potato, Dextrose, Agar medium) slant. Culture tubes are incubated at 25±2°C temperature for eight days. If a luxuriant growth of any fungus on a paper/cloth / wooden object is observed, it can be directly transferred with the help of an inoculation needle in culture tubes containing PDA slants.

If the growth of fungi appearing on the object is of different colour, it should be carefully isolated into different culture tubes, because there is always a possibility of more than one fungus present at the site.

Experiments on the effect of hydrogen ion concentration indicates that pH 5.5 is optimum for both growth and sporulation in all the organisms studied. Therefore, initial pH of the medium is adjusted to 5.5 by adding 0.1 N HCl/NaOH solution in all cultural studies. pH of the medium is determined with the help of a digital pH meter.

Adjustment of temperature :

Optimum temperature for growth and sporulation is found 25±2°C and during different cultural studies this temperature was maintained.

Another method of isolating the fungus is to keep the material in a moist or humid place for a few days and isolate the fungus by scrapping. Since both these acts may prove harmful, these methods are not advisable for museum objects.

3.3 Study of aeromycoflora

Different methods have been suggested to isolate the indoor aeromycoflora or the fungal organisms present in the environment and many of these are potential biodeteriogens. Two methods are commonly used -

1. *Culture plate exposure method*:

This method was first used by Hyde and Williams (1961). This method has an advantage that we can easily identify the colony produced by different fungi. However, if a non-viable fungal spore or mycelium is present it will not be recorded.

This method includes exposure of petridish containing culture medium for a recorded time. To get better results, more than one petridishes containing PDA are used as replicates. Similarly more than one media can also be used. The height at which petridishes are kept, the medium used and the time of exposure will determine the appearance of colonies after 5 to 7 days. The petridishes are incubated at 25±2°C and an early observation is required, so that the members of Mucorales may not over grow in the petridishes.

2. *Gravity slide exposure method:*

This method includes the suction of air by artificial means and slides coated with glycerine jelly are placed in the chamber. The fungal spores present in the air settle down and are trapped by glycerine jelly coated slides. By taking repeated samples or replicates, the desired qualitative and quantitative results are obtained. It is also possible to identify certain fungi by staining the slides with cotton blue but to differentiate spores of same morphology is a major drawback. All the spores which are trapped may be viable or non viable spores. Therefore, a total count can be recorded by this method. Different air samplers have been designed to obtain such data.

3.4 Purification of Fungal Cultures:

Before proceeding for identification it is essential to purify the cultures. Most of the time during isolation a mixed culture is obtained, i.e. a petridish or culture tube having two or more fungi or a presence of bacteria and fungus together. The methods for purification of cultures are as follows:

(a) Purification by altering the pH:

The best way to separate a fungus from a bacterium is to make the pH of the medium acidic. At pH 5 or below 5, the growth of bacteria will be very much reduced (Brown 1924) and a pure fungal colony thus can be separated.

(b) Purification by sub-culturing :

When a mixed culture is obtained it is transferred to a petridish containing PDA medium. After 3-4 days when growth initiation takes place a small piece of PDA alongwith growing tips of hyphae is transferred to another tube. Such process is repeated if required. This technique is based on the principle that different fungi have different growth rates. The fungus growing fast in PDA thus can be isolated in a pure form.

(c) Single spore culture or hyphal tip method:

The cultures obtained by above two methods are in strict sense not pure cultures. These are termed as a axenic cultures.

To obtain a single spore culture the fungus is grown in PDA. When good sporulation is observed the spore suspension may be obtained by spreading sterile distilled water over the fungal growth. This suspension, in a very dilute form is spread on a petridish and with the help of student microscope a single spore or hypha is located. It is then cut by Dummy cutter objective (La Rua cutter devised by Keyworth, 1959). Thus obtained pure culture may be maintained on PDA slants and can be used for cultural or control studies. Inoculation is carried out under asceptic conditions in a Horizontal Laminar Flow (Yorco Horizontal Laminar Flow, Type No. YSL-188).

3.5 Identification of Fungi:

For identification of organisms morphological and microscopic observations are required. Visual observation can tell that a culture belongs to a fungus or a bacterium. Bacterial culture will always be watery or give shining appearance. In **fungal cultures** cottony or mycelial growth can be observed. If a culture tube is filled with fungal mycelium i.e. the growth is luxuriant the fungi may belong to group Mucorales.

The colour of colony is helpful in identifying species of *Aspergillus* and *Penicillium*. The colour of the backside of the agar can be used as a criterion to identify various fungi like *Fusarium, Curvularia* and *Alternaria*. Visual observations give a clue but the microscopic observations actually give the base for identifying numerous biodeteriogens.

The structure of mycelium, septate or unseptate, the structure of conidia, their colour, size, shape and structure of fruiting bodies are helpful; to identify a fungus.

The correct identification of fungal cultures can be made by sending the pure culture tubes to any reputed fungal culture collections. Some of these are listed below:

1. Indian Type Culture Collection; Division of Mycology and Plant Pathology, IARI, New Delhi 110012.
2. Microbial type culture collection & Gene Bank, Institute of Microbial Technology, 1389 Sector 33C, Chandigarh 160031.
3. Fungal Culture Collection, Agharkar Research Institute, Pune.
4. International Mycological Institute (IMI) Bakeham Lane, Egham,Surrey TW 20 9TY, U.K.
5. American Type Culture Collection (ATCC) 12301 Parklawn Drive Rockville, Maryland 20852, U.S.A.
6. Collection of Institute Pasteur (CIP) 25-28, rue du Docteur Roux 75724 Paris Cedex 15, France.
7. Czechoslovak Collection of Mciroorganism, J.E. Purkyne University, Tr. Obrancu miru 10, Prague, Czechoslovakia.
8. Japan Collection of Microorganisms, The Institute of Physical and Chemical Research, Japan.

9. Deutsche Sammlung von Mikroorganismen, Grise bachstrasse 8 D-3400 Gottingen, Fedral Republic of Germany.

3.6 Assessment of Fungal Growth and Biochemical Studies:

There are two methods to assess the growth of any fungus:

(a) **Colony diameter method :**

In the first method a diameter of a fungal colony can be measured directly from the back of the petridish in mm after a certain period.

(b) **Dry weight method:**

In the second method a fungal inoculum is to be incorporated in a basal medium and then to be incubated at 25$\pm$2°C temperature for required days. The fungal mat is harvested at the end of the incubation period by filtering over previously dried and weighed Whatman filter paper No. 1. Mat is then thoroughly washed with double distilled water and subsequently dried in an electric oven at 65°C for 50 hours. Dried mats with filter papers are transferred to a desiccator to bring down to room temperature. The filter paper containing fungal mat is rapidly weighed on an analytical balance. Difference of the weights of the filter papers can be calculated and thus weight of the fungal mat can be known.

In order to study the effect of various compounds containing carbon, nitrogen, sulphur and phosphorous, the amount of individual substance in a basal medium is calculated and quantity equivalent to that should be substituted singly to the basal medium by replacing the original corresponding substance viz. glucose, potassium nitrate, magnesium sulphate and potassium dihydrogen phosphate respectively. The amount of polysaccharides should be kept similar to that of glucose. Medium devoid of glucose, potassium nitrate, magnesium sulphate and potassium dihydrogen phosphate will serve as control for carbon, nitrogen, sulphur and phosphorus respectively in Asthana and Hawker's medium 'A'.

Chromatographic Studies:

Chromatographic studies are useful in understanding utilization pattern of different mono, oligo and polysaccharides. Solutions containing mono saccharides are to be autoclaved at 15 lbs p.s.i for 20 minutes while fractional sterilization is to be carried out for di and polysaccharides. Filtrate is to be analysed each day chromatographically to detect the presence of these sugars or other sugars produced during the growth.

The paper chromatograms are run in n-butanol-acetic acid-water (4:1:5

v/v; Ranjan *et al.* 1955), and they are subsequently dried at room temperature. In order to separate glucose and galactose n-butanol-pyridine-water (6:4:3 v/v) is used as a solvent. A mixture of n-butanol, aniline-phthalic acid and distilled water (100 ml n-butanol+10 ml distilled water +0.91 ml aniline+1.6 g phthalic acid) is used as a spraying reagent. Sprayed chromatograms are again dried and then placed in an electric oven at 110oC for 90 second, to develop the characteristic bands (Fig. 1). Rf values of various sugars are recorded. Presence of starch can be determined with the help of potassium iodide solution (Ghosh and Tandon, 1965).

3.7 Biodeterioration of Wood and Paper :

Biodeterioration of wood and paper is studied by taking known amount of their scrap in 150 ml Erlenmeyer conical flask. 25 ml of modified Asthana and Hawker's medium 'A' was used as a basal medium. Flasks are autoclaved as mentioned earlier and then organisms were inoculated in these flasks. They were incubated at 25$\pm$2°C for 25 days. The fungal mat was then harvested along with remaining wood or paper by filtering over previously dried and weighed Whatman's filter paper No. 1. Actual weight of fungal mat can be deducted to obtain actual loss of wood or paper during the incubation period.

3.8 Control of Biodeteriogens

Control of the selected organisms was studied *in vitro* by using poisoned food technique of Nene and Thapliyal (1979). A known amount of chemical / fungicide is added in PDA medium before plating and then organisms are inoculated. To observe the effect of natural plant products aqueous extracts are prepared by crushing 100 g leaf in 100 ml water. The extract is centrifuged and water is added latter on to make the desired concentrations. Oils are diluted by mixing suitable solvents. The results are recorded after a definite time, i.e., 7 days at 25$\pm$2°C temperature in the present investigation.

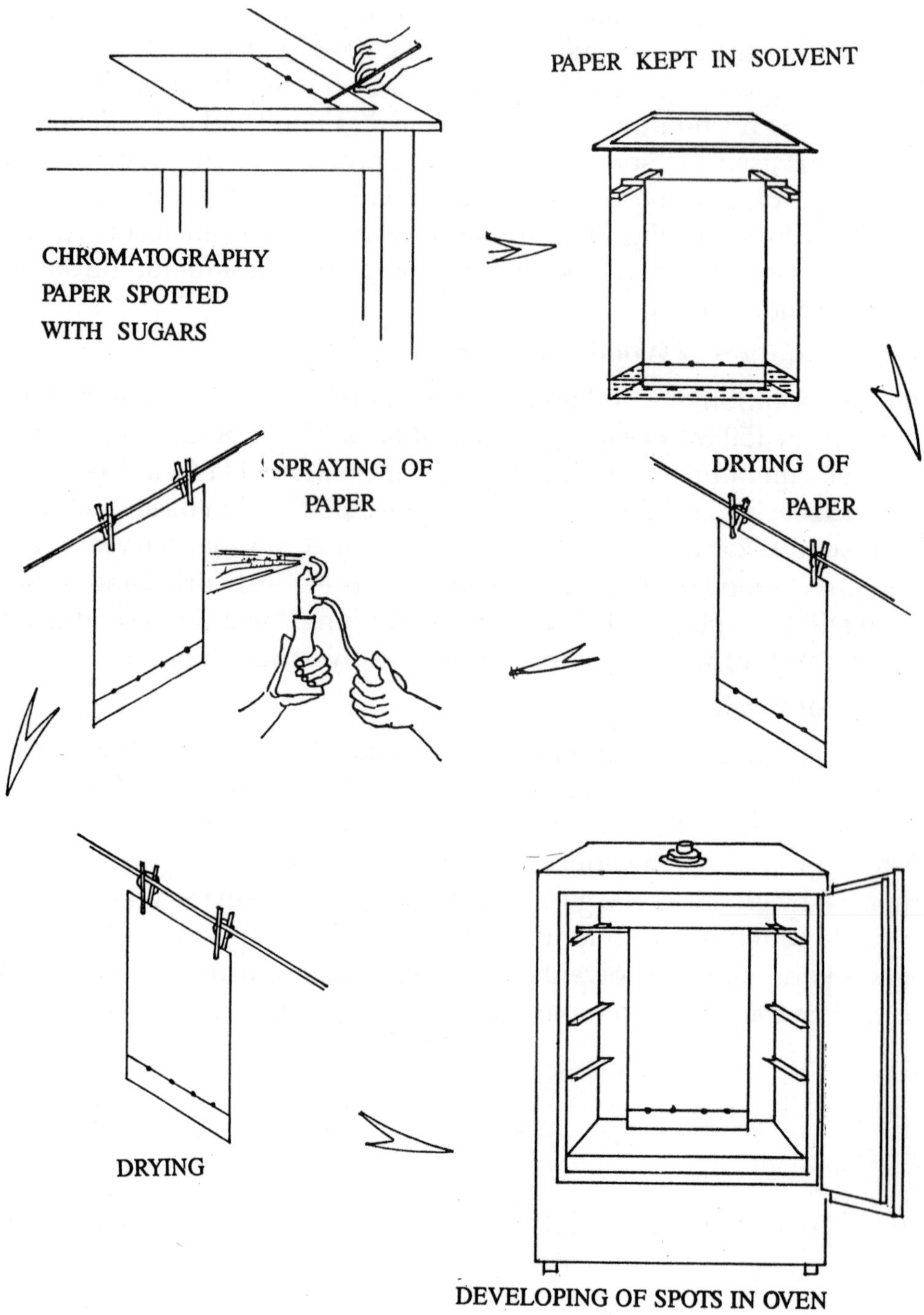

Fig. 1 : Technique of paper chromatography

Chapter – 4

Isolation Studies

Fungi constitute a major group of organisms which may cause severe and irreparable damage to cultural property. It will not only cause destruction but sometimes it may produce very dirty stains which will disfigure the object. Control of fungi attacking on museum objects is possible only if a systematic study of individual fungus has been made. The behaviour of individual organism and their metabolic processes must be understood. A study of nutritional requirement of an organism is also very important. But to study all these things, isolation of fungi from museum objects is essential. During the present study of different museum objects were found to be affected by various fungi. A list of fungi isolated from museum objects is given in tables 1 & 2.

First of all the affected objects were taken out from the exhibition, to avoid infection to other materials. Subsequently they were cleaned with absolute alcohol in order to remove fluffy growth of fungi. After some days when fungal growth reappeared, it was isolated from individual objects on PDA slants (Fig. 2). Cultures were incubated for 7 days at the temperature 25$\pm$2°C. On the basis of morphological characters and microscopic examination fungi were identified. They were finally confirmed by the IMI, Kew, England.

Table 1: List of fungi isolated from Museum objects.

Date of isolation	*Name of the object (material)*	*Place*	*Organisms*
20-7-90	Wooden plaque (wood)	Department of Museology, M. S. University, Baroda.	*Aspergillus niger* van Teighem
21-8-91	Pavri- A (wood)	Department of Museology, M. S. University, Baroda.	*Aspergillus Japonicus* K. Saito *A. flavus* Link *Penicillium citrinum* Thom
21-8-91	Pavri-B (wood)	Department of Museology, M. S. University, Baroda.	*Cochliobolus pallescens* (Tsuda and Ueyama) Sivan *Phoma* sp. *Fusarium semitectum* Berk. & Rav.

Date of isolation	*Name of the object (material)*	*Place*	*Organisms*
17-9-91	Parsi Manuscript Book (Paper)	Department of Museology, M. S. University, Baroda.	*Cochliobolus lunatus* R. R. Nelson & F. A. Haasis *Chaetomium cymbiforme* Lodha *Aspergillus niger* van Tieghem A. *oryzae* (Ahlburg) Cohn. A. *fumigatus* Fresen.
7-10-91	Flute (Wood)	Department of Museology, M. S. University, Baroda.	*Cochliobolus verruculosus* (Tsuda and Ueyama) Sivan *Penicillium citrinum* Thom.
9-10-91	Tambura (Wood)	Department of Museology, M. S. University, Baroda.	*Emericella nidulance* (Eldam) Vuill.
9-7-92	Miniature Painting (Paper)	Department of Museology, M. S. University, Baroda.	*Aspergillus tamarii.* Kita *Cladosporium herbarum* (Pers.) Link.
25-9-91	Miniature Painting (Paper)	Museum and Picture Gallery, Baroda.	*Aspergillus terreus*. Thom A. *versicolor* (Vuill) Tiraboschi *Chaetomium gangligerum* L. Ames
10-10-91	Chinese Wooden Screen (Wood)	Museum and Picture Gallery, Baroda.	*Aspergillus ustus* (Bainier) Thom. and Church *Aspergillus nidulans* (Eidam) Winter
12-12-92	Photograph B/W (Paper)	Department of Museology, M. S. University, Baroda.	*Aspergillus flavus* Link *Syncephalastrum racemosum* Cohn ex Schroter
21-7-93	Lace work (Textile)	Museum and Picture Gallery, Baroda.	*Aspergillus* sp. *Aspergillus oryzae* (Ahlburg) Cohn. *A. flavus* Link
3-10-93	Photograph B/W (Paper)	Museum and Picture Gallery, Baroda.	*Chaetomium* sp. *Trichoderma kuningii* Oudem
4-12-93	Musical Instruments 1. Transu 2. Khanjari 3. Dilruba (Leather)	Kutch Museum, Bhuj	*Aspergillus flavus* Link

Table 2 : List of fungi isolated from museum allied objects

Date of isolation	*Name of the object (material)*	*Place*	*Organisms*
17-9-91	Leather	Private collector Baroda	*Penicillium citrinum* Thom. *Aspergillus niger* van Tieghem
20-7-92	Wooden Art Piece (Wood)	Private collector Baroda	*Trichoderma* sp. *Geotrichum* sp.
6-8-92	Panther Skin	Private collector Baroda	*Aspergillus niger* van Tieghem
17-2-93	Canvas Painting (Textile)	Private collector Baroda	*Aspergillus niger* van Tieghem *Rhizopus* sp.
17-5-93	Book (Paper)	Personal library Baroda	*Aspergillus niger* van Tieghem
12-5-93	Khesh (Textile)	Private collector Baroda	*Aspergillus fumigatus* Fresen *Absidia spinosa* Lendner.
12-5-93	Manuscript (Paper)	Private collector Baroda	*Ascotricha erinaceae* Berk.

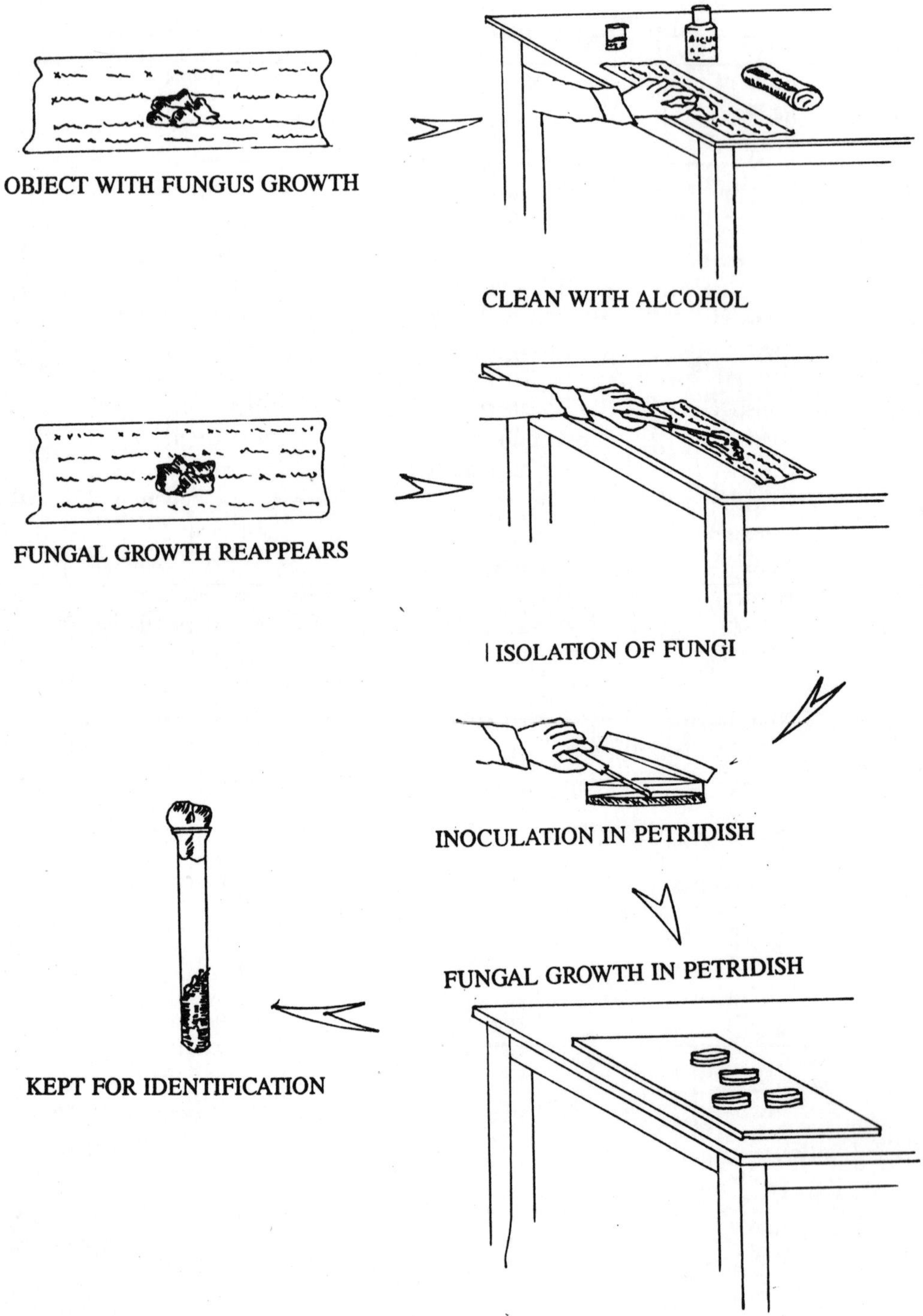

Fig. 2 : Different steps to isolate the fungal biodeteriogens.

Isolations were made from :—

(1) Museum of the Department of Museology, M. S. University, Baroda;

(2) Museum and Picture Gallery, Baroda;

(3) Kutch Museum, Bhuj and

(4) Private collectors.

Six species of *Aspergillus*, three species of *Cochliobolus* and one species each of *Penicillium, Phoma, Fusarium, Chaetomium,* and *Cladosporium* were found during isolation of fungi in the Department of Museology, M. S. University, Baroda. Occurrence of *Aspergillus* and *Penicillium* on museum objects is common but the rest of the fungi have not been reported frequently.

The reason for fungal growth on these objects is the prevailing high percentage of relative humidity (RH) during monsoon season. A measurement of temperature and relative humidity was taken which revealed the presence of about 90% RH during monsoon season. As this museum is situated near the bank of the river Vishwamitri, RH is always found somewhat more as compared to other locations in the city. The museum of the Department of Museology came into existance in January 1990. Prior to that objects were found lying in the store. This may be the another reason for the growth of fungi. However, these objects are not damaged much as the fungus growth was noticed in a very early stage. Only some disfigurement due to stains was observed. Stains were removed by chemical treatment and fumigation of the infected objects was done in an airtight chamber, using antifungal chemicals as a fumigant.

Table 1 shows that certain objects like the miniature painting, Chinese wooden screen, textile and photographs were found affected by fungi in the Museum and Picture Gallery, Baroda.

The reason for fungal attack may be prevailing high RH inside the museum building during rainy season. The wooden Chinese screen which was displayed on the wall in 'The Greater India Room' was damaged due to luxurient fungal growth. It was due to the seepage of water from the roof. The wooden screen was decorated with precious stones fixed on lacquer layer. Due to the physiological processes and metabolic products of the fungi, lacquer layer started separating from its base and hence precious stones also started to fall down. Many of the precious stones had separated and lacquer layer had also started cracking and flaking. The isolation of the fungi was done in the laboratory of the Museum and Picture Gallery. Two fungi were identified. These were *Aspergillus ustus* and *A. nidulans.*

Kutch Museum, Bhuj has a good collection of musical instruments. During the visit to Bjuj Museum, It was noticed that some musical instruments were affected by fungi. Fungal growth was found on the leather part of the musical instruments. The fungi were isolated with the help of a piece of cellotape. Fungal spores were transferred on the piece of cellotape and it was kept in separate petridishes. These fungal spores were grown on PDA medium in the laboratory and then they were identified. Only *A. flavus* was found present on all the three objects.

Several objects belonging to private collectors were also found affected by fungi (Table 2). These objects were of varied nature such as wooden object, animal skin, canvas painting, textile object, paper material and leather etc. Staining was the major problem of these objects. The fungal attack may be due to the unscientific storage of the objects. The isolated fungi included species of *Aspergillus* and *Penicillium.* The occurrence of *Geotrichum, Absidia,* and *Ascotricha* is not common on such objects.

All types of objects like books, paper print, palm leaves, leather puppets, leather objects, textiles, paintings etc., are affected in a museums (Jeyaraj 1991). In India, a variety of writing materials have been used since ancient times. In Orissa and South India, palm leaf has been used for writing religious texts, epic poems etc., from very early times. In regions of high altitude,mainly in the Northern parts of India, birch bark was the writing material. For ordinary correspondence, some cheap material which was easily perishable was used. Important documents like religious edicts and legal documents were incised on strong material like stone and copper etc. (Bhatnagar 1991). Dhawan (1986) reported 35 different fungi affecting paper. Nyuksha (1994) found development of 288 species belonging to 167 genera on paper. Indian miniature paintings probably constitute one of the richest treasures of Indian cultural heritage. Biodeterioration is primarily due to favourable climatic conditions and the constituent materials of the paintings (Uniyal 1991). 23 fungal species were recorded from 800 miniature paintings of State Museum Lucknow (Dhawan and Agrawal 1986).

Chapter–5

Cultural Studies

5.1 Selection of Suitable Culture Medium

"There is no universal natural substrate or artificial medium upon which all fungi will grow", (Lilly and Barnett, 1951). Even closely related forms may differ considerably in their nutritional requirements qualitatively as well as quantitatively.

Arya (1982), Rai (1982) and Mathew (1993) have reported modified Asthana and Hawker's medium 'A' suitable for the growth of different fungal organisms.

According to Lilly and Barnett (1951), 'In selecting a medium the purpose for which it is to be used should be kept in view. For many purposes a natural medium is one of the choice'. A natural medium is a medium which is made up of natural materials of unknown composition. They further advocated for the selection of a synthetic medium. The essentials of a synthetic medium may be stated as follows : Source of C = Carbon, N = Nitrogen in utilizable form, P = Phosphate and S = Sulphate ions and others.

Considering the growth of fungi, in a particular habitat, attention is always to be given to the source of nutrition. This is particularly important where highly recalcitrant substrates are concerned. The actual quantity of food material required to support mould growth is very small. Fungi depend upon their environment for energy sources. Carbon, Nitrogen, Sulphur, Phosphorus, Sodium, Potassium, Calcium, Magnesium, Iron, Manganese, Zinc, Copper, Cobalt, Molybdenum and Vitamins are some of the important components for the growth of an organism.

A knowledge of nutrition of the fungi is necessary for culturing them in laboratory or in industry. In the beginning cultivation of fungi was unintentional, later-on developed into an art in connection with the preparation of food stuffs and beverages. Natural materials were used in the early periods for culturing fungi in laboratories. During last century many workers have tried to develop semisynthetic as well as synthetic media to cultivate fungi in laboratories.

In order to select a suitable basal medium for fungi under the present investigation, different media were tried and their mycelial growth and sporulation were recorded.

Figure-3 depicts the combination of seven different media used in the present study and the results are summarised in tables 3 to 6 and in figure-4.

It is evident from the tables 3 to 6 that Richard's medium supported best growth of *Cochliobolus verruculosus* followed by Czapek's medium, Dox's medium, Modified Asthana and Hawker's medium 'A', Elliott's medium, Asthana and Hawker's medium 'A' and Coon's medium.

In the case of *P. citrinum* growth was maximum on Richard's medium followed by Czapek's, Dox's medium, Modified Asthana and Hawker's medium 'A', Asthana and Hawker's medium 'A', Elliott's and Coon's media.

DOX'S MEDIUM	
Sucrose	15.00g
$NaNO_3$	2.00g
KH_2P0_4	1.00g
$MgSO_4 7H_2O$	0.05g
KCl	0.05g
$FeSO_4 7H_2O$	Trace
Distilled Water	1000ml

ASTHANA AND HAWKER'S MEDIUM 'A'	
D Glucose	5.00g
KNO_3	3.50g
KH_2PO_4	1.75g
$MgSO_4 7H_2O$	0.75g
Distilled Water	1000ml

CZAPEK'S MEDIUM	
Sucrose	3.00g
$NaNO_3$	2.00g
KH_2PO_4	1.00g
KCl	0.50g
$MgSO_4 7H_2O$	0.50g
$FeSO_4 7H_2O$	0.01g
Distilled Water	1000ml

ELLIOTT'S MEDIUM	
D Glucose	5.00g
L-Asparagine	1.00g
KH_2PO_4	1.36g
Na_2CO_3	1.06g
$MgSO_4 7H_2O$	0.50g
Distilled Water	1000ml

COON'S MEDIUM	
Maltose	3.50g
L-Asparagine	0.25g
KH_2PO_4	1.25g
$MgSO_4 7H_2O$	0.50g
Distilled Water	1000ml

RICHARD'S MEDIUM	
Sucrose	50.00g
KNO_3	10.00g
KH_2PO_4	5.00g
$MgSO_4 7H_2O$	2.50g
$FeCl_3$	0.02g
Distilled Water	1000ml

MODIFIED ASTHANA AND HAWKER'S MEDIUM 'A'	
D Glucose	10.00g
KNO_3	3.50g
KH_2PO_4	1.75g
$MgSO_4 7H_2O$	0.75g
Distilled Water	1000ml

Fig. 3 : Composition of different nutritional media tried

Table 3: Average dry weight (in mg), sporulation and change in pH by *Cochliobolus verruculosus* in different culture media.

Sl. No.	*Media*	*Dry Weight*	*Sporulation*	*Final pH*
1	Dox's Medium	220	Poor	7.0
2	Asthana and Hawker's medium 'A'	025	Fair	7.0
3	Coon's Medium	025	Poor	5.5
4	Czapek's Medium	350	Excellent	3.5
5	Richard's Medium	413	Excellent	7.0
6	Elliott's Medium	030	Poor	6.5
7	Modified Asthana and Hawker's Medium 'A'	185	Good	6.0
	G.M.	169.7		

Summary of dry weight results and conclusion at 5% level of p.

Treatments : Highly significant
Replicates : Non-significant
S.E. : 4.7542
C.D. at 5% level : ±10.1979
Dry weight results : 5 > 4 > 1 >7 >$\overline{6\ 2\ 3}$

Table-4 : Average dry weight (in mg), sporulation and change in pH by *Penicillium citrinum* in different culture media.

Sl. No.	*Media*	*Dry Weight*	*Sporulation*	*Final pH*
1	Dox's Medium	115	Excellent	6.5
2	Asthana and Hawker's Medium 'A'	050	Good	6.0
3	Coon's Medium	003	Nil	5.5
4	Czapek's Medium	205	Excellent	3.5
5	Richard's Medium	351	Excellent	6.5
6	Elliott's Medium	022	Fair	6.5
7	Modified Asthana and Hawker's Medium 'A'	099	Good	6.0
	G.M.	120.7		

Summary of dry weight results and conclusion at 5% level of p.

Treatments : Highly significant
Replicates : Non-significant
S.E. : 5.7679
C.D. at 5% level : ±12.3723
Dry weight results : 5>4>1>7>2>6>3

Table-5 : Average dry weight (in mg), sporulation and change in pH by *Aspergillus ustus* in different culture media.

Sl. No.	*Media*	*Dry Weight*	*Sporulation*	*Final pH*
1	Dox's Medium	105	Excellent	8.0
2	Asthana and Hawker's Medium 'A'	025	Good	7.5
3	Coon's Medium	003	Nil	5.5
4	Czapek's Medium	235	Excellent	7.5
5	Richard's Medium	220	Excellent	9.0
6	Elliott's Medium	011	Poor	7.0
7	Modified Asthana and Hawker's Medium 'A'	090	Good	5.5
	G.M.	98.43		

Summary of dry weight results and conclusion at 5% level of p.

Treatments : Highly significant
Replicates : Non-significant
S.E. : 10.6413
C.D. at 5% level : ±22.8256
Dry weight results : $4>5>\overline{1\ 7}>\overline{2\ 6\ 3}$

Table-6 : Average dry weight (in mg), sporulation and change in pH by *Chaetomium cymbiforme* in different culture media.

Sl. No.	*Media*	*Dry Weight*	*Sporulation*	*Final pH*
1	Dox's Medium	125	Excellent	7.0
2	Asthana and Hawker's Medium 'A'	032	Fair	7.C
3	Coon's Medium	010	Nil	5.5
4	Czapek's Medium	300	Excellent	7.0
5	Richard's Medium	365	Excellent	6.5
6	Elliott's Medium	031	Poor	6.5
7	Modified Asthana and Hawker's Medium 'A'	110	Good	7.5
	G.M.	139.0		

Summary of dry weight results and conclusion at 5% level of p.

Treatments : Highly significant
Replicates : Non-significant
S.E. : 4.8827
C.D. at 5% level : ±10.4735
Dry weight results : $5>4>1>7>\overline{2\ 6}>3$

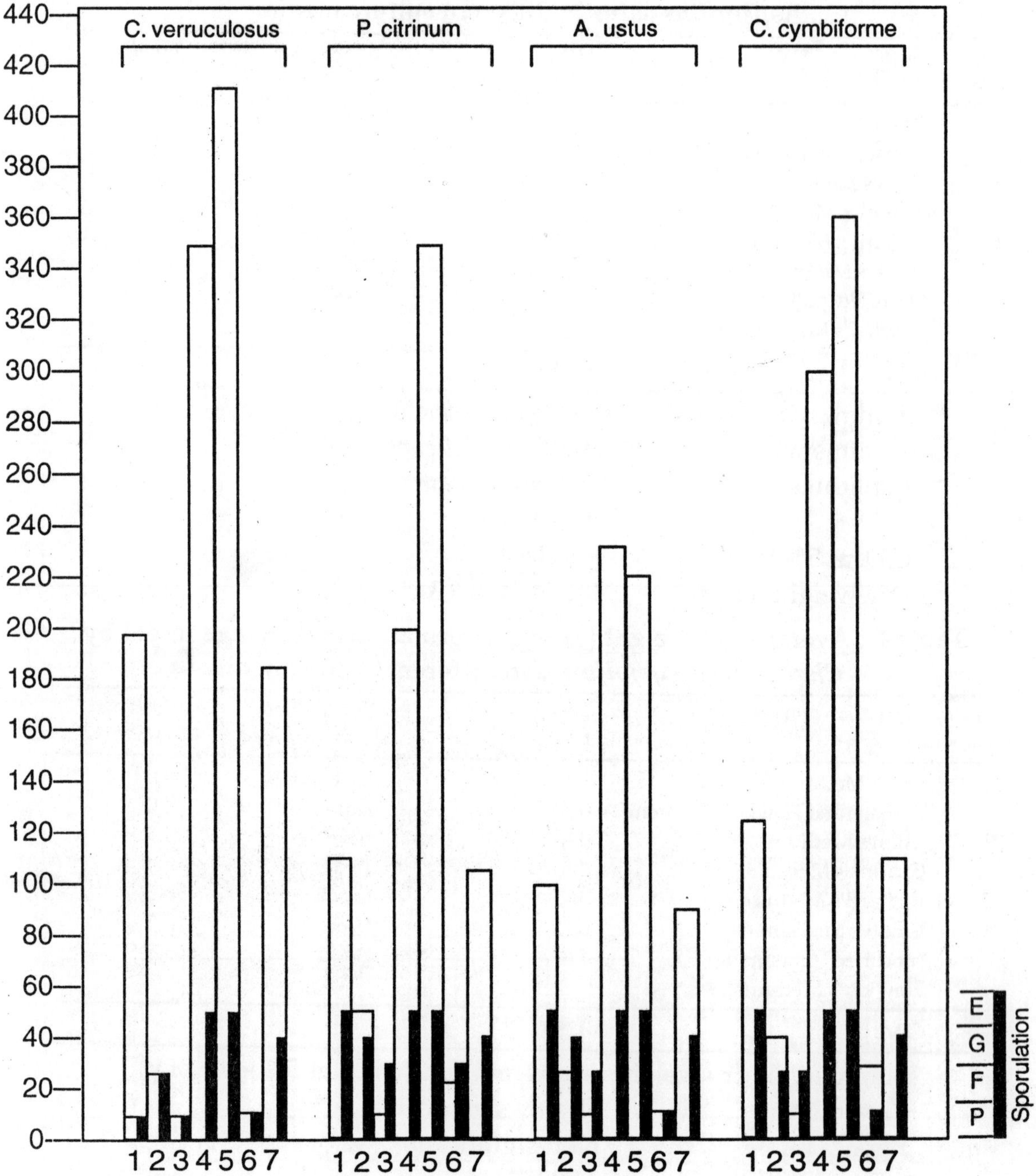

Fig. 4 : Average dry weight and sporulation of *C. verruculosus, P. citrinum, A. ustus, C.cymbiforme* in different culture media.

The growth of *A. ustus* was maximum in Czapek's followed by Richard's, Dox's medium, Modified Asthana and Hawker's medium 'A', Asthana and Hawker's medium 'A', Elliott's and Coon's medium.

Dry weight of *Chaetomium cymbiforme* was maximum in Richard's followed by Czapek's, Dox's medium, Modified Asthana and Hawker's medium 'A', Asthana and Hawker's medium 'A', Elliott's and Coon's medium.

Excellent sporulation was found in all the four fungi in Richard's and Czapek's medium, while good sporulation was found in modified Asthana and Hawker's medium 'A'. Poor to fair sporulation was observed in Elliott's, Coon's and Asthana and Hawker's and Dox's medium.

At the end of incubation period pH of the medium shifted from neutral to alkaline side. However, pH drift was towards acidic range in Czapek's medium in *C. verruculosus* and *P. citrinum.* There was no change in pH of Coon's medium in all the organisms.

Among the seven media tried, Richard's medium showed the best growth of all the fungi under present study except *A. ustus.* It showed the best growth in Czapek's medium.

Although modified Asthana and Hawker's medium 'A' did not support excellent growth statistically, compared to Richard's medium, it supported sufficient growth and sporulation in all the cases. Moreover, this medium is easy to manipulate with respect to the expected need for modifications and substitutions of its constituents. Hence, it is used in all subsequent cultural studies.

5.2 Influence of Different Hydrogen Ion Concentrations.

Microorganisms are highly sensitive to hydrogen ion concentrations of the media in which they grow. Most of the commonly occuring moulds tolerate a wide range of hydrogen ion concentration, provided other conditions are favourable. Some prefer acidic media, others neutral and some others grow best in alkaline conditions. In general, slightly acidic medium is favourable to spore germination and rapid growth of the young colony. Once the growth is established, the reaction of the medium usually changes, due to accumulation of products of metabolism (Pathak and Dhawan, 1989). Hydrogen ion concentration affects permiability of protoplasmic membranes, uptake of minerals entry of essential vitamins and organic acids into the cells, activity of enzyme systems, synthesis and stability of proteins and other similar life processes.

The pH of the medium also affects the processes like production of pigments, vitamins and other metabolites.

The upper and lower pH values between which a fungus is able to grow is called the pH range. The pH ranges of various species are different. Fungi generally tolerate more acid than alkali. The optimum pH range may be different for growth and sporulation. The pH of a medium in which a fungus is growing may change. High buffer concentration and limited growth may keep the changes in pH at a minimum.

Four metabolic processes operate to change the pH of a culture medium:

(1) Utilization of cations, (2) Utilization of anions, (3) Formation of acids from neutral metabolites (especially carbohydrates), and (4) Formation of bases (especially ammonia) from amino acids and proteins. The net change in pH is the result of the interaction of all these processes.

The organic catalysts have great influence on growth and sporulation patterns of an organism. It has been found by various mycologists that, enzymes have optima between pH 4 and 8. According to Lilly and Barnett (1951), "The chemical changes in media due to alteration of pH, whether imposed from outside or caused by the fungus, affects metabolic processes".

The rise in pH of the culture medium has been attributed to the metabolic activities during growth resulting in absorption of anion or production of ammonia from nitrogenous compounds. Lowering of pH in case of media with higher initial pH was possible due to absorption of carbon dioxide produced by the fungus during the process of respiration (Singh, 1977).

Yu (1954) reported good growth over a range of pH from 3 to 12 in *Ascobolus magnificus*. *Phyllosticta cycadina* grew between pH 3 and 9 (Bilgrami, 1956). *Botryodioplodia theobromae* and *Hendersonula toruloidea* could tolerate a pH range of 2.0 to 12.0 (Williamson, 1964). A slightly narrower pH range (3.0 to 7.2) was observed for *Phoma herbarum* (Chung, 1967). Ross (1960) reported that *Fusarium culmarum, F. sambucinum* var. *coeruleum and F. oxysporum* grew well between pH 5 and 8. Bhargava (1962) found the optimum range for the growth of *Fusarium solani* as 4.5 to 6.5. Malca *et al.* (1966) as well as Selvaraj (1971) reported that higher pH values upto 10.0 are better tolerated in *Verticillium* and *Fusarium* spp. Lakshmikant *et al.* (1990) have reported that *Chaetomium globosum* showed maximum degradation of cellulose at pH 5.

In the light of the above findings, it was decided to study the effect of pH of the medium on the growth and sporulation of four fungi under present study. The initial pH values of modified Asthana and Hawker's medium 'A' were adjusted from 2.0 to 10.0 (i. e. 2.0, 3.0, 3.5, 4.0, 4.5, 5.0, 5.5, 6.0, 6.5, 7.0, 8.0, 9.0 and 10.0). Results obtained are recorded in tables 7 to 10 and in fig. 5.

Results indicate that all the four organisms could grow and sporulate over a wide range of pH. *C. verruculosus* and *C. cymbiforme* failed to grow at pH 2.0 like *Phomopsis gulabia* and *P. pedilanthi* (Arya, 1990-91), while other two organisms showed minor growth at the same pH. Contrary to it *Phomopsis viticola* and *P. psidii* grew between pH 2-10 (Arya, 1991). On the acidic side all the four organisms showed significantly good growth rather than in alkaline medium. With the increase in acidity upto 5.5, dry weight of the organisms also showed increase while it decreased at the pH 5.5 similar to *P. viticola* (Arya, 1991). Arya (1990) studied the influence of H^+ concentration on the utilization of sodium nitrite by four different spp. of *Phomopsis*. He reported that none of the organisms was able to grow between pH 2 and 5.5. A considerable growth was recorded at pH 6.0 in all the cases.

In *C. verruculosus* and *C. cymbiforme* sporulation was found good in acidic media upto 5.5, after which it was fair to poor. In *P. citrinum* and *A. ustus* sporulation was found good from pH 2 to 9.

It was observed that the pH of the medium of all the four fungi drifted towards neutral point or slightly alkaline side at the end of incubation period.

The results of the experiments showed that the best growth and sporulation was observed at pH 5.5 and hence in all the subsequent experiments the initial pH was adjusted to 5.5.

Table-7 : Average dry weight (in mg), sporulation and final pH of *Cochliobolus verruculosus* at various pH values

Sl. No.	*pH of the Medium*	*Dry Weight*	*Sporulation*	*Final pH*	*Drift in pH*
1	2.0	00	Nil	2.0	0.0
2	3.0	066	Good	7.0	+4.0
3	3.5	070	Good	7.0	+3.5
4	4.0	058	Good	7.5	+3.5
5	4.5	060	Fair	7.5	+3.0
6	5.0	095	Good	6.0	+1.0
7	5.5	126	Good	6.0	+0.5
8	6.0	065	Fair	8.5	+2.5
9	6.5	046	Nil	8.0	+1.5
10	7.0	045	Poor	9.0	+2.0
11	8.0	035	Nil	8.0	0.0
12	9.0	040	Poor	7.5	-1.5
13	10.0	030	Nil	8.0	-2.0
	G.M.	56.61			

Summary of dry weight results and conclusion at 5% level of P.

Treatments	:	Highly significant.
Replicates	:	Non-significant.
S.E.	:	9.7944
C.D. at 5% level	:	±20.1374
Dry weight result	:	7>6>3 2 8 5 4 9 10 12 11 13 > 1

Table-8 : Average dry weight (in mg), sporulation and final pH of *Penicillium citrinum* at various pH values.

Sl. No.	*pH of the Medium*	*Dry Weight*	*Sporulation*	*Final pH*	*Drift in pH*
1	2.0	048	Good	7.0	+5.0
2	3.0	076	Good	6.5	+3.5
3	3.5	058	Good	6.5	+3.0
4	4.0	054	Good	7.5	+3.5
5	4.5	058	Good	4.5	+3.0
6	5.0	085	Good	6.0	0.0
7	5.5	099	Good	6.0	+1.0
8	6.0	063	Poor	7.0	+0.5
9	6.5	079	Poor	7.5	+1.0
10	7.0	046	Fair	6.5	+1.0
11	8.0	038	Good	8.0	-0.5
12	9.0	045	Good	6.5	-2.5
13	10.0	030	Fair	7.0	-3.0
	G.M.	60.00			

Summary of dry weight results and conclusion at 5% level of P.

Treatments	:	Highly significant.
Replicates	:	Non-significant.
S.E.	:	4.2102
C.D. at 5% level	:	±8.6563
Dry weight result	:	7> 6> 9 2> 8 3 5 4 1 10 12 11 13

Table-9 : Average dry weight (in mg), sporulation and final pH of *Aspergillus ustus* at various pH values.

Sl. No.	*pH of the Medium*	*Dry Weight*	*Sporulation*	*Final pH*	*Drift in pH*
1	2.0	042	Good	6.0	+4.0
2	3.0	055	Good	7.5	+4.5
3	3.5	055	Good	7.5	+4.0
4	4.0	060	Good	8.0	+4.0
5	4.5	062	Good	7.5	+3.0
6	5.0	065	Good	6.5	+1.5
7	5.5	083	Good	5.5	0.0
8	6.0	062	Good	7.5	+1.5
9	6.5	059	Fair	8.0	+1.5
10	7.0	050	Good	8.0	+1.5
11	8.0	050	Good	8.0	0.0
12	9.0	043	Good	8.0	-1.0
13	10.0	031	Fair	8.0	-2.0
	G.M.	55.15			

Summary of dry weight results and conclusion at 5% level of P.

Treatments	:	Highly significant.
Replicates	:	Non-significant.
S.E.	:	4.8286
C.D. at 5% level	:	±9.9277
Dry weight result	:	7 > 6 5 8 4 9 3 2 10 11 12 1 > 13

Table - 10 : Average dry weight (in mg), sporulation and final pH of *Chaetomium cymbiforme* at various pH values

Sl. No.	*pH of the Medium*	*Dry Weight*	*Sporulation*	*Final pH*	*Drift in pH*
1	2.0	00	Nil	2.0	0.0
2	3.0	045	Poor	7.5	+4.5
3	3.5	045	Poor	7.5	+4.0
4	4.0	050	Fair	7.5	+3.5
5	4.5	065	Fair	7.5	+3.0
6	5.0	087	Good	7.0	+2.0
7	5.5	105	Good	7.5	+2.0
8	6.0	079	Good	7.5	+1.5
9	6.5	069	Fair	7.5	+1.0
10	7.0	063	Fair	7.5	+0.5
11	8.0	054	Fair	7.5	-0.5
12	9.0	044	Fair	7.0	-2.0
13	10.0	033	Fair	8.0	-2.0
	G.M.	56.84			

Summary of dry weight results and conclusion at 5% level of P.

Treatments	:	Highly significant.
Replicates	:	Non-significant.
S.E.	:	3.02129
C.D. at 5% level	:	± 6.2117
Dry weight result	:	7 > 6 > 8 > 9 5 10 > 11 4 3 2 12 > 13 > 1

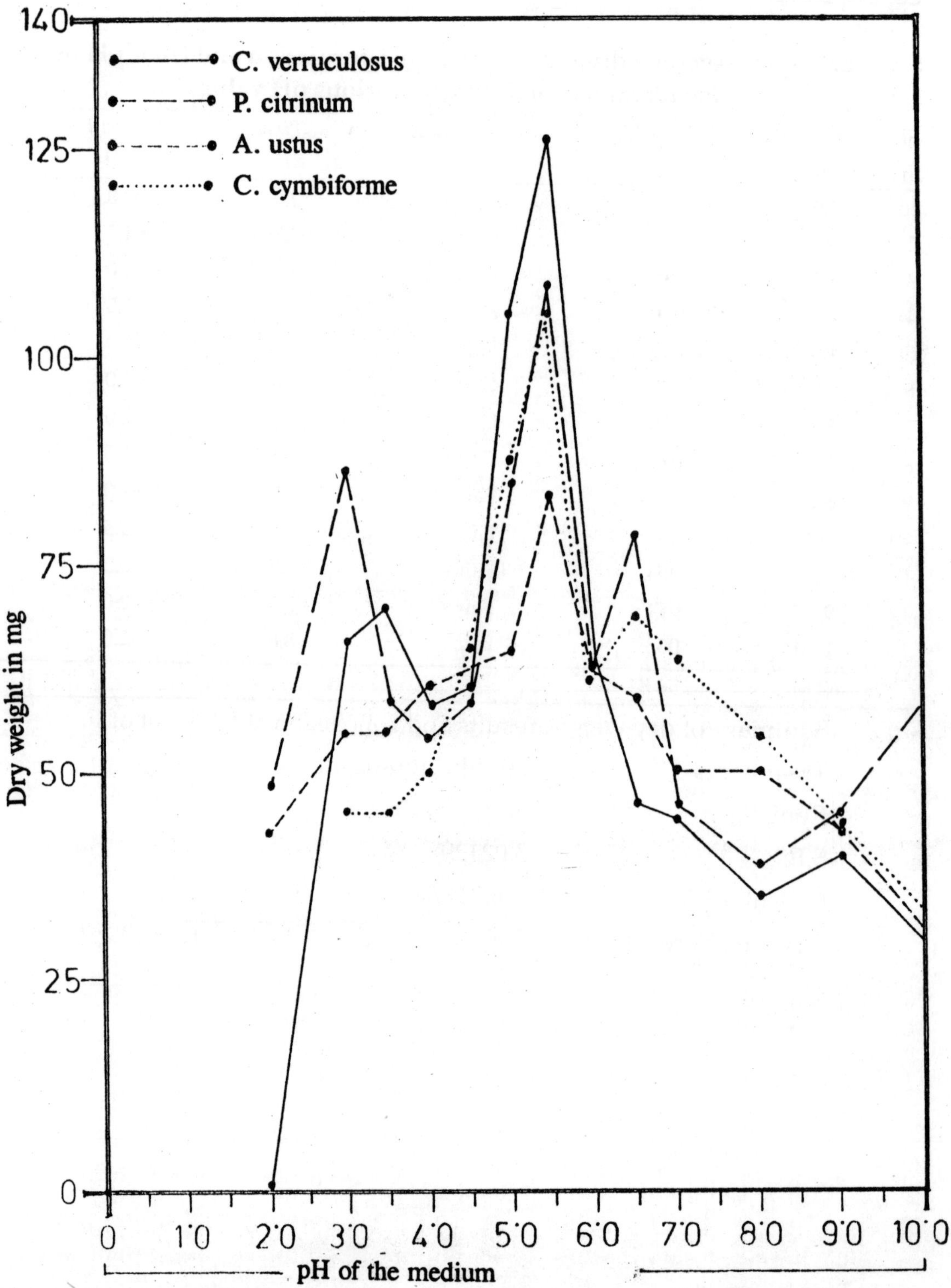

Fig. 5 : Effect of different hydrogen ion concentrations on growth of four fungi.

5.3 Influence of Different Temperatures

Temperature has a great influence on the growth and sporulation of an organism. The temperature determines various metabolic activities of fungi and even distribution of pathogenic fungi around different parts of the world. In the same way temperature is considered as one of the most important environmental factors for the growth of fungi on museum materials. The range of temperature, which favours mycelial growth varies considerably. It depends upon the organism concerned. According to Deverall (1965), "Temperature affects all the fungal cellular activities and it also causes shift in metabolism." Usually fungi do not grow below 0°C and above 40°C. However, certain species are reported growing even below freezing point and above 40°C. The optimum temperature for fungal growth is between 20°C and 30°C.

Agrawal and Dhawan (1985) reported that fungi can tolerate extreme temperature range from 10° to 80°C (or even more or less) but rapid growth can be observed between 25° and 35°C temperature.

Each species has its own particular temperature at which it grows best. That particular temperature will be 'optimum temperature' for that fungal species.

Cochrane (1958) stated that spores of *Cladosporium* sp. and *Sporotrichum* sp. germinated in cold storage at temperatures below freezing point. The lowest temperature range for fungal growth is 0°C to 7°C for *Penicillium* sp. (Mislivee and Tuite, 1970).

Kingsland (1985) reported that *Aspergillus flavus*, being thermotolerant, is able to tolerate even a temperature as high as 110°C for 30 min.

The temperature influences growth by affecting the enzyme system; low temperature generally slows down metabolism, while higher temperature increases the rate of cell activities. Every microorganism has certain range of lower and higher temperature beyond which growth ceases.

Based on the limits of temperature within which they grow, fungi may be divided into three major groups :

(1) Thermophiles :

These fungi can tolerate temperature from 20° to 50°C. Fungi tolerating high temperature but growing at low temperature are called thermotolerant fungi. Thermophilic fungi are generally present in places exposed to higher solar radiations.

(2) Psychrophiles :

The optimum growth of these fungi is obtained below 20°C.

(3) Mesophiles :

These fungi are in between thermophilic and psychrophilic fungi. Most of the fungi are mesophilic. They grow actively at temperatures between 10° and 40°C.

The optimum range of temperature for the growth of fungi on museum material is 15° to 30°C (Agrawal & Pathak). Optimum temperature permitting growth may not be always similar to that permitting good sporulation. Cochrane (1958) concluded, "only two generalizations can be formulated with respect to temperature and sporulation:"

1. The temperature range permitting reproduction is usually narrower than that permitting growth.

2. Each spore form has its own temperature optimum, which may or may not coincide with that of vegetative growth or of other spore forms of the same species."

The above mentioned facts show that it is necessary to have a thorough knowledge about the temperature requirement of the test organisms, before starting any physiological studies. This would also give us an idea about the most suitable environmental conditions for the survival of these fungi. Therefore, it was decided to determine the cardinal temperature for the growth and sporulation of the fungi under study.

The organisms were exposed to different temperatures such as 5, 10 15, 20, 25, 30, 35, 40, 45, 50 and 55°C. To test the survival of these organisms at lower and higher temperatures, the inoculated flasks were brought back to the room temperature (25±2°C). Increase in fungal mat after seven days showed the survival of the fungus at that temperature. The results are recorded in tables 11 to 14 and shown in fig. 6.

The results depicted in the tables 11 to 14 show that the maximum growth occurs at the temperature 25°C in all the four organisms similar to *Colletotrichum dematium* and *C. gloeosporioides* (Rajak, 1983). However, Gonzaler *et al.* (1988) and Montani *et al.* (1989) reported optimum growth of *P. citrinum* at a temperature of 30°C in Czapek's agar medium. Singh (1980) studied the effect of temperature on growth behaviour of *Alternaria brassicae* and *Drechslera graminea* and found optimum growth of these organism at 22° and 24°C respectively. *Pleurotus membranaceous* and *Sarcoxylon companctum,* two wood rot fungi, grew best at 24° and 28°C respectively (Sarwal *et al.*, 1982). Luque (1989) reported optimum temperature range 25°-28°C for *Diplodia mutila*. At 5°C *C. verruculosus* showed some growth while rest of the organisms failed to grow at this temperature, however, the spores were found alive upto 15 days. It is clear that growth of the organisms accelerated with an increase in temperature upto 25°C. But at the higher temperature it decreased in all the organisms. The mycelial dry weight of the organisms, depicted in tables 11 to 14 shows the rise or fall with increase in temperature. At 45°C *C. verruculosus* could not grow and its spores were found dead, while, *P. citrinum* also could not grow at 45°C, but its spores were found alive at this temp. On the other hand *A. ustus* has shown growth at 50°C whereas, *C. cymbiforme* has shown growth upto 45°C. *A. ustus* could not grow at the temperature of 55°C. Spores of both the organisms were found dead at 55°C.

Table - 11: Average dry weight (in mg), sporulation and final pH of *Cochliobolus verruculosus* at different temperatures.

Sl. No.	*Temperature in °C*	*Dry Weight (mg)*	*Sporul-ation*	*Final pH*	*Remarks*
1	5	40	Poor	6.0	
2	10	44	Poor	5.5	
3	15	56	Poor	5.5	
4	20	155	Fair	7.0	
5	25	185	Good	7.0	
6	30	161	Fair	7.0	
7	35	126	Fair	7.2	
8	40	80	Nil	5.0	
9	45	00	Nil	5.0	Spores found dead
10	50	00	–	–	
11	55	00	–	–	
	G.M.	94.11			

Summary of dry weight results and conclusion at 5% level of P.

Treatments	:	Highly significant.
Replicates	:	Non-significant.
S.E.	:	4.3688
C.D. at 5% level	:	±8.7419
Dry weight result	:	$5 > \overline{6\ 4} > 7 > 8 > 3 > 2 > 1 > \overline{9\ 10\ 11}$

Table - 12 : Average dry weight (in mg), sporulation and final pH of *Penicillium citrinum* at different temperatures.

Sl. No.	*Temperature in °C*	*Dry Weight (mg)*	*Sporul-ation*	*Final pH*	*Remarks*
1	5	00	Nil	-	Spores found alive
2	10	00	Nil	-	Spores found alive
3	15	71	Poor	5.5	
4	20	76	Good	7.0	
5	25	99	Good	7.0	
6	30	80	Good	7.5	
7	35	77	Good	7.0	
8	40	53	Good	5.0	
9	45	00	Nil	5.0	Spores found alive
10	50	00	Nil	5.0	Spores found dead
11	55	00	-	-	
	G.M.	50.66			

Summary of dry weight results and conclusion at 5% level of P.

Treatments	:	Highly significant.
Replicates	:	Non-significant.
S.E.	:	2.1372
C.D. at 5% level	:	$\pm$4.2766
Dry weight result	:	$5 > \overline{6\ 7\ 4} > 3 > 8 > \overline{1\ 2\ 9\ 10\ 11}$

Table - 13 : Average dry weight (in mg), sporulation and final pH of *Aspergillus ustus* at different temperatures.

Sl. No.	*Temperature in °C*	*Dry Weight (mg)*	*Sporul-ation*	*Final pH*	*Remarks*
1	5	00	Nil	-	Spores found alive
2	10	00	Nil	5.5	Spores found alive
3	15	50	Nil	5.5	
4	20	62	Nil	7.0	
5	25	90	Good	7.0	
6	30	83	Excellent	7.0	
7	35	68	Excellent	7.5	
8	40	53	Excellent	7.0	
9	45	50	Good	7.0	
10	50	76	Good	7.0	
11	55	00	Nil	-	Spores found dead
	G.M.	48.36			

Summary of dry weight results and conclusion at 5% level of P.

Treatments	:	Highly significant.
Replicates	:	Non-significant.
S.E.	:	5.983
C.D. at 5% level	:	±12.4090
Dry weight result	:	$\overline{5\ 6} > \overline{10\ 7} > \overline{4\ 8\ 3\ 9} > \overline{1\ 2\ 11}$

Table - 14 : Average dry weight (in mg), sporulation and final pH of *Chaetomium cymbiforme* at different temperatures.

Sl. No.	*Temperature in °C*	*Dry Weight (mg)*	*Sporul-ation*	*Final pH*	*Remarks*
1	5	00	Nil	5.5	Spores found alive
2	10	00	Nil	5.5	Spores found alive
3	15	45	Nil	5.5	
4	20	97	Good	7.0	
5	25	110	Good	7.0	
6	30	105	Excellent	7.5	
7	35	105	Good	8.0	
8	40	87	Fair	8.0	
9	45	70	Fair	8.0	
10	50	00	Nil	5.0	Spores found alive
11	55	00	-	-	Spores found dead
	G.M.	56.30			

Summary of dry weight results and conclusion at 5% level of P.

Treatment	:	Highly significant.
Replicates	:	Non-significant.
S.E.	:	1.7320
C.D. at 5% level	:	$\pm$ 3.6129
Dry weight result	:	$5 > \overline{6\ 7} > 4 > 8 > 9 > 3 > \overline{1\ 2\ 10\ 11}$

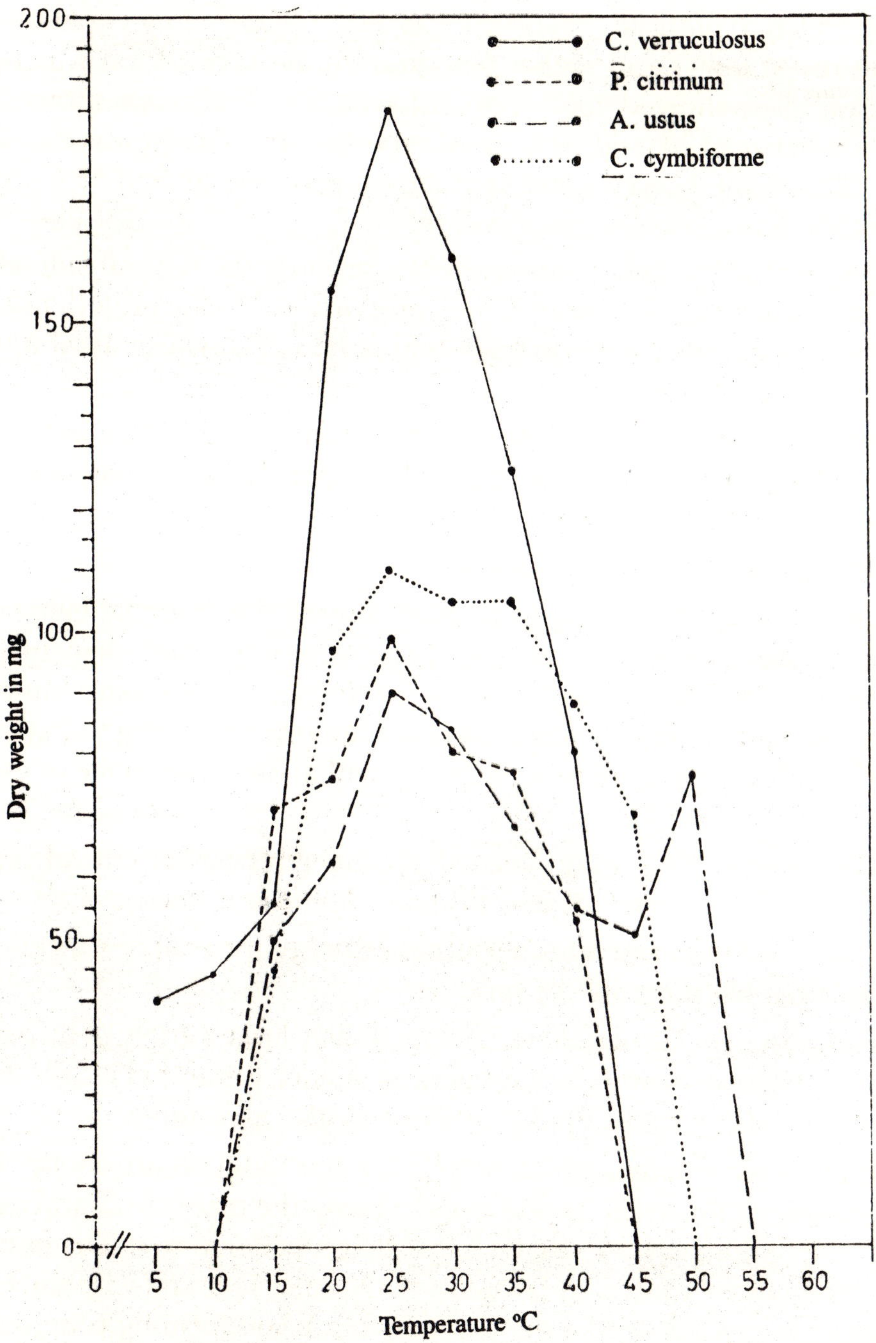

Fig. 6 : Effect of different temperatures on four fungi.

Sporulation of these organisms was also influenced by variation in the temperature. At lower and higher temperature the sporulation was poor to nil. *C. verruculosus* showed good sporulation at 25°C. *P. citrinum* showed good sporulation between 20° and 40°C. In *A. ustus* the sporulation was excellent at 30°C. *C. cymbiforme* showed good sporulation between 20° and 35°C. At 30°C the sporulation was excellent in this case.

The final pH of the medium drifted towards the neutral side after 15 days of incubation period. The drift was proportional to the growth rate. In *A. ustus* and *C. cymbiforme* there was a gradual drift in pH towards alkalinity with increase in temperature.

Since the results showed by all the four fungi at 25°C are indicative of good mycelial growth and good or excellent sporulation, it was selected for all subsequent cultural studies.

5.4 Effect of Different Carbon Sources.

Carbon occupies a unique position among the essential elements required by living organisms. It is one of the most important elements in the fungal nutrition and has to play a pivotal role in their metabolism. Almost half of the dry weight of fungus cells consists of carbon. According to Pathak and Dhawan (1989), carbon sources serve two essential functions in the physiology of fungi.

1. They supply the carbon needed for the synthesis of critical constituents, such as carbohydrates, proteins, lipids and nucleic acids.
2. Their oxidation provides a source of energy for proper functioning of essential life processes of fungi.

Fungi are capable of utilizing tremendously large variety of carbon compounds for the growth and development as sugars, organic acids etc. to large polymers such as proteins, lipids, polysaccharides and lignin.

It is stated by Lilly and Barnett (1951) that fungi secure energy by oxidizing organic compounds. In addition, to being the main structural element, carbon compounds play an equally important functional role. The number of carbon compounds known so far exceeds the total known compounds of all the other elements because of the property of carbon forming compounds in which carbon is linked to carbon in the form of chains and rings.

According to Chander *et al.* (1977) glucose is the most effective source of carbon among all the tested carbohydrates for *Penicillium chrysogenum*. Fungi can also utilize carbon sources other than carbohydrates, such as organic acids, amino acids and their derivatives, some polycyclic compounds and alkaloids. *Aspergillus nidulans* can use ethanol as a carbon source for growth and conidiation, and is more effective than glucose (Martinelle, 1976). Not only the type but also the concentration of the carbon source is important to determine its effectiveness in promoting growth.

The carbohydrates are the first choice of a fungus. Whether a sugar is utilized or not, depends upon both, the configuration of the sugar and the particular abilities of the specific fungus. Relatively, the simpler carbohydrates are utilized more efficiently, while the complex ones are successfully used by those fungi only, which possesses necessary enzymes to cause their disintegration (Arya, 1982).

According to Lilly and Barnett (1951) following generalization about utilization of the common hexoses may be made:

1. There is no single sugar which supports the maximum amount of growth for all of these fungi;
2. All of these fungi utilize glucose, although the maximum amount of growth was not always attained on this sugar and;
3. The more closely the configuration of another sugar approaches that of glucose, the more fungi utilize it.

Among hexoses, D-glucose and D-fructose are very efficiently used by all the fungi. D-glucose is biologically the most important and utilized for growth virtually by all cultivable fungi.

Energy and most of the cell building materials for the wood rotting fungi are supplied by carbohydrate fraction (Garg and Mukerji 1994). Cellulose, hemicellulose and lignin comprise more than 95% of the substance of most woods. Water soluble nutrients can be moved through wood and get deposited near the surface of cut timber in normal drying. These may help to colonize the wood. Cellulose is the chief nutrient present in paper.

In order to understand the physiology of the fungal organisms, a knowledge of carbon nutrition is essential. Therefore, it was decided to study the influence of different carbon sources on the growth and sporulation of four fungi under study. Following 10 carbon sources were used and results are recorded in tables 15 to 18 and in figs. 7 to 10.

1. **Carbohydrates:**
   (a) Monosaccharides:
      (i) D-glucose
      (ii) D-fructose
   (b) Disaccharides :
      (i) Maltose
      (ii) Sucrose
      (iii) Lactose
   (c) Polysaccharides :
      (i) Soluble starch
      (ii) Inulin
      (iii) Carboxy Methyl Cellulose (C.M.C.)
2. **Organic acids :**
   (a) Dicarboxylic acid
      Oxalic acid
   (b) Tricarboxylic acid
      Citric acid

Monosaccharides :

These are simple sugars having general formula $C_n (H_2O)_n$. The carbon chain is unbranched except in a few, very rare sugars. The hexoses such as D-glucose and D-fructose occur naturally. They are sweet in taste and soluble in water.

Disaccharides :

These sugars are derived from two hexose sugars by the elimination of water. On hydrolysis, the individual sugars are regenerated.

Polysaccharides :

The chemistry of the polysaccharides resembles to that of the oligosaccharides except that the number of sugar residues is much larger. These substances constitute the bulk of carbohydrate materials synthesized by plants and animals. The most important polysaccharides are cellulose, starch and glycogen. On hydrolysis simple sugars are formed. In general, polysaccharides are insoluble or only colloidally soluble.

Organic Acids :

An organic acid is characterized by having one or more carboxyl (-COOH) groups. Some organic acids are utilized as a sources of carbon and in other ways.

Results shown in tables 15 to 18 and figs. 7 to 10 indicate that all the four fungi under present study were capable to grow in all the carbon sources except oxalic acid. But the amount of growth varied in individual carbon sources. Carbon was found indispensable for the growth because none of the fungi could grow in its absence.

In the present investigation D-glucose provided an excellent growth in all the four fungi. Mathew (1993) reported excellent growth of *Fusarium pallidoroseum* and *F. oxysporum* on this source. Arya (1991) also reported good growth in all the four species of *Phomopsis*. Saini & Sidhu (1983) reported best growth of *Choanephora cucurbitarum* on glucose. Bilgrami & Prasad (1979) reported that increase in mycelial growth of *Cochliobolus specifer* was proportional to glucose utilized. The fungus utilized glucose more rapidly than fructose.

Sporulation was found good in all the four fungi on D-glucose. Similar observations were made by Mathew (1993) for *Fusarium pallidoroseum, F. oxysporum* and *F. moniliforme*.

Table - 15 : Average dry weight (in mg), sporulation and final pH of *C. verruculosus* on different carbon sources.

Sl. No.	*Carbon Source*	*Dry Weight (mg)*	*Sporulation*	*Final pH*
1	D-Glucose	185	Good	6.5
2	D-Fructose	107	Good	7.5
3	Maltose	120	Poor	7.5
4	Sucrose	231	Poor	6.0
5	Lactose	071	Poor	7.0
6	Soluble starch	151	Nil	7.5
7	Inulin	089	Fair	8.0
8	C.M.C.	272	Nil	6.0
9	Oxalic acid	000	Nil	5.5
10	Citric acid	070	Good	11.0
11	No Carbon	000	Nil	5.5
	G.M.	117		

Summary of dry weight results and conclusion at 5% level of P.

Treatments	:	Highly significant.
Replicates	:	Non-significant.
S.E.	:	3.8058
C.D. at 5% level of p	:	±7.8934
Dry weight result	:	$8 > 4 > 1 > 6 > 3 > 2 > 7 > \overline{5\ 10} > \overline{9\ 11}$

Table - 16 : Average dry weight (in mg), sporulation and final pH of *P. citrinum* on different carbon sources.

Sl. No.	*Carbon Source*	*Dry Weight (mg)*	*Sporulation*	*Final pH*
1	D-Glucose	099	Good	6.0
2	D-Fructose	099	Good	7.5
3	Maltose	078	Excellent	7.0
4	Sucrose	097	Excellent	7.0
5	Lactose	078	Good	6.0
6	Soluble starch	085	Excellent	6.0
7	Inulin	095	Good	7.5
8	C.M.C.	095	Nil	6.0
9	Oxalic acid	000	Nil	5.5
10	Citric acid	076	Poor	11.0
11	No Carbon	000	Nil	5.5
	G.M.	073		

Summary of dry weight results and conclusion at 5% level of P.

Treatments	:	Highly significant.
Replicates	:	Non-significant.
S.E.	:	8.3690
C.D. at 5% level of P	:	$\pm$17.3573
Dry weight result	:	1 2 4 7 8 6 3 5 10 > 9 11

Table - 17 : Average dry weight (in mg), sporulation and final pH of *A. ustus* on different carbon sources.

Sl. No.	*Carbon Source*	*Dry Weight (mg)*	*Sporulation*	*Final pH*
1	D-Glucose	090	Good	5.5
2	D-Fructose	077	Excellent	8.0
3	Maltose	081	Good	5.5
4	Sucrose	086	Excellent	7.0
5	Lactose	054	Good	7.5
6	Soluble starch	086	Excellent	7.5
7	Inulin	100	Fair	7.0
8	C.M.C.	122	Nil	6.0
9	Oxalic acid	000	Nil	5.5
10	Citric acid	077	Good	11.0
11	No Carbon	000	Nil	5.5
	G.M.	070.3		

Summary of dry weight results and conclusion at 5% level of P.

Treatments	:	Highly significant.
Replicates	:	Non-significant.
S.E.	:	6.4432
C.D. at 5% level of P	:	±13.3632
Dry weight result	:	$8 > \overline{7\ \underline{1}}\ \underline{4\ 6\ 3\ 2\ 10} > 5 > \overline{9\ 11}$

Table-18 : Average dry weight (in mg), sporulation and final pH of *C. cymbiforme* on different carbon sources.

Sl. No.	*Carbon Source*	*Dry Weight (mg)*	*Sporulation*	*Final pH*
1	D-Glucose	110	Good	7.5
2	D-Fructose	092	Good	8.0
3	Maltose	094	Fair	7.0
4	Sucrose	128	Excellent	7.5
5	Lactose	097	Good	7.0
6	Soluble starch	107	Excellent	7.5
7	Inulin	106	Excellent	7.5
8	C.M.C.	044	Nil	6.0
9	Oxalic acid	000	Nil	5.5
10	Citric acid	041	Poor	8.0
11	No Carbon	000	Nil	5.5
	G.M.	074.4		

Summary of dry weight results and conclusion at 5% level of P.

Treatments	:	Highly significant.
Replicates	:	Non-significant.
S.E.	:	8.8887
C.D. at 5% level of P	:	±18.4352
Dry weight result	:	4 1 6 7 5 3 2 > 8 10 > 9 11

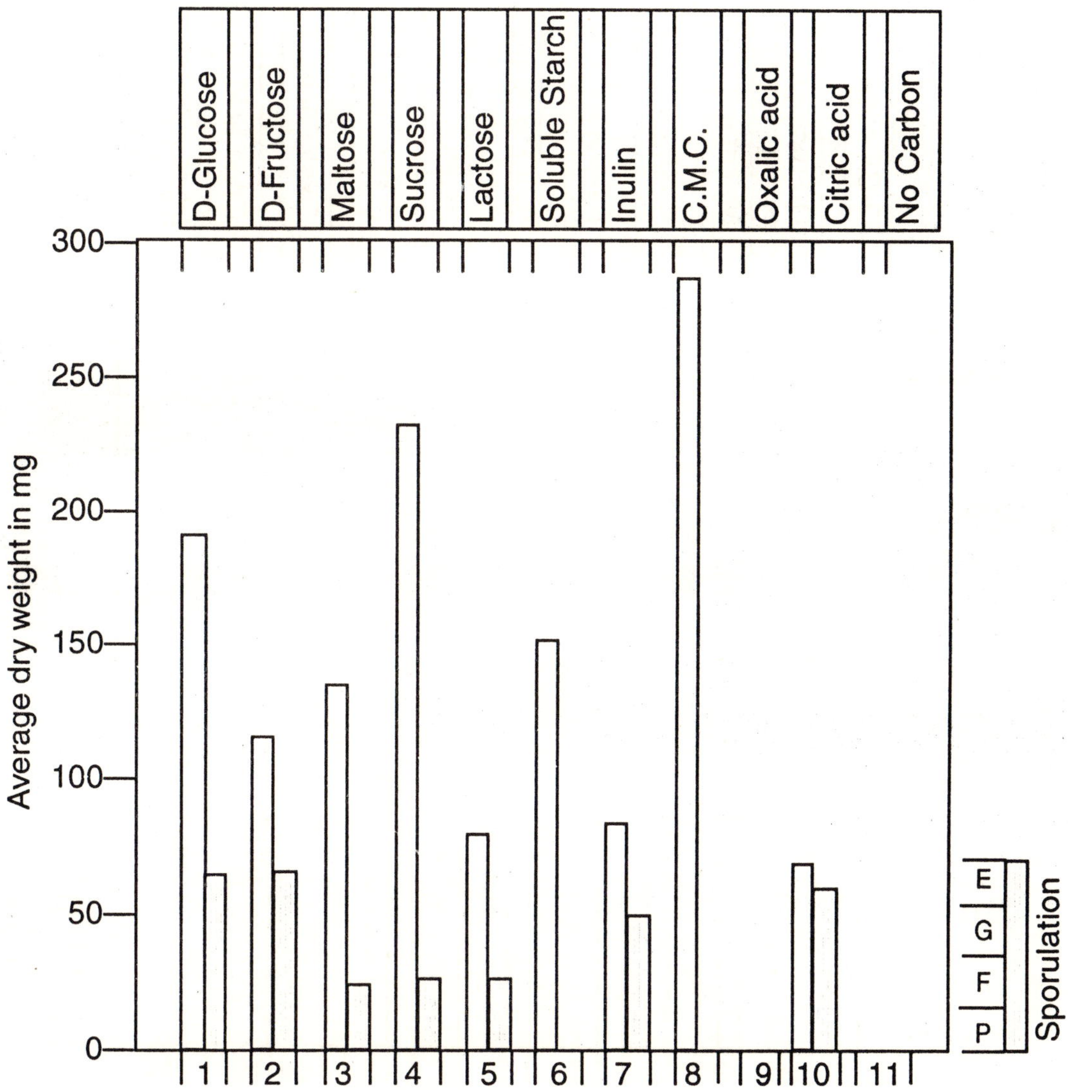

Fig. 7 : Average dry weight and sporulation of *C. verruculosus* on various Carbon sources.

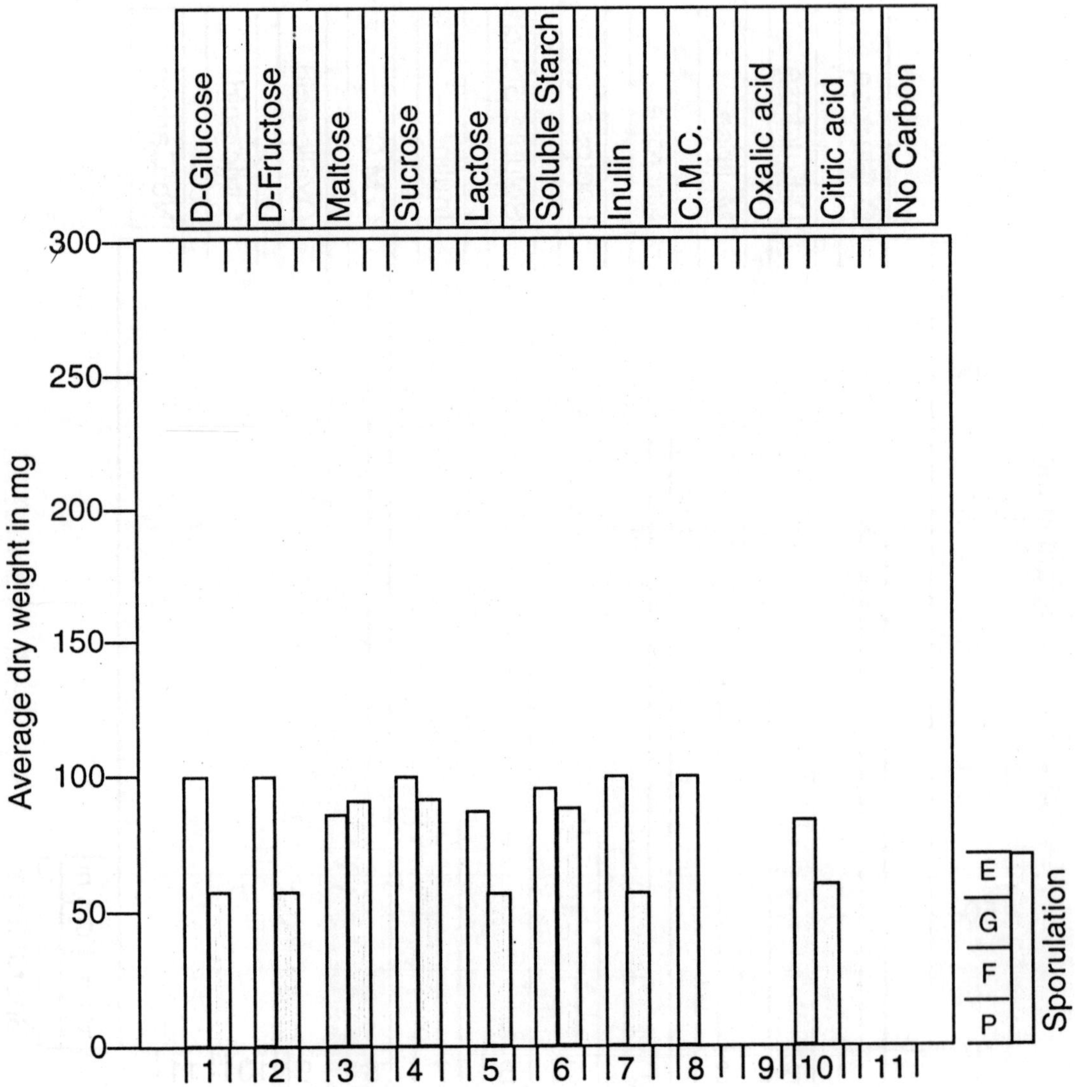

Fig. 8 : Average dry weight and sporulation of *P. citrinum* on various carbon sources.

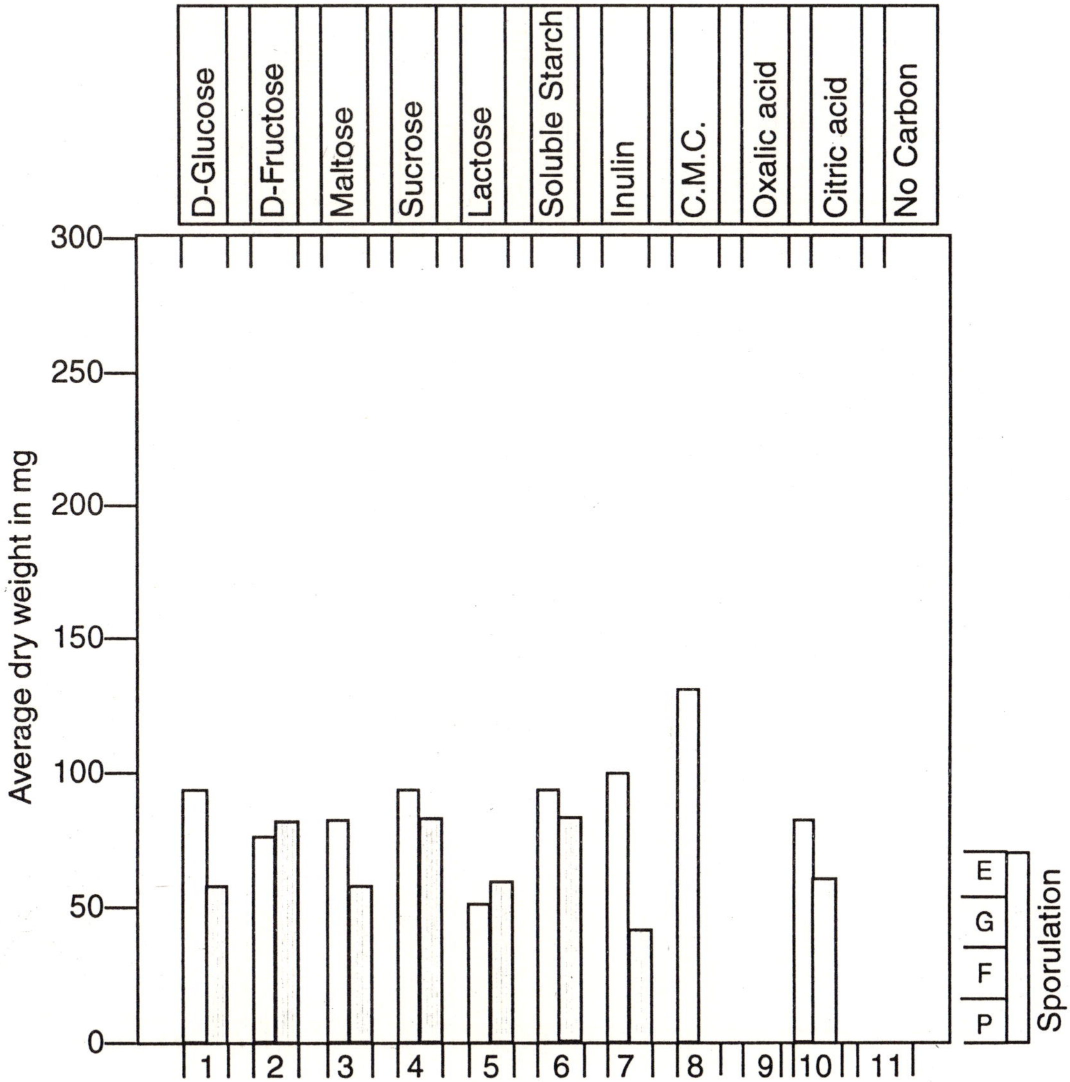

Fig. 9 : Average dry weight and sporulation of *A. ustus* on various carbon sources.

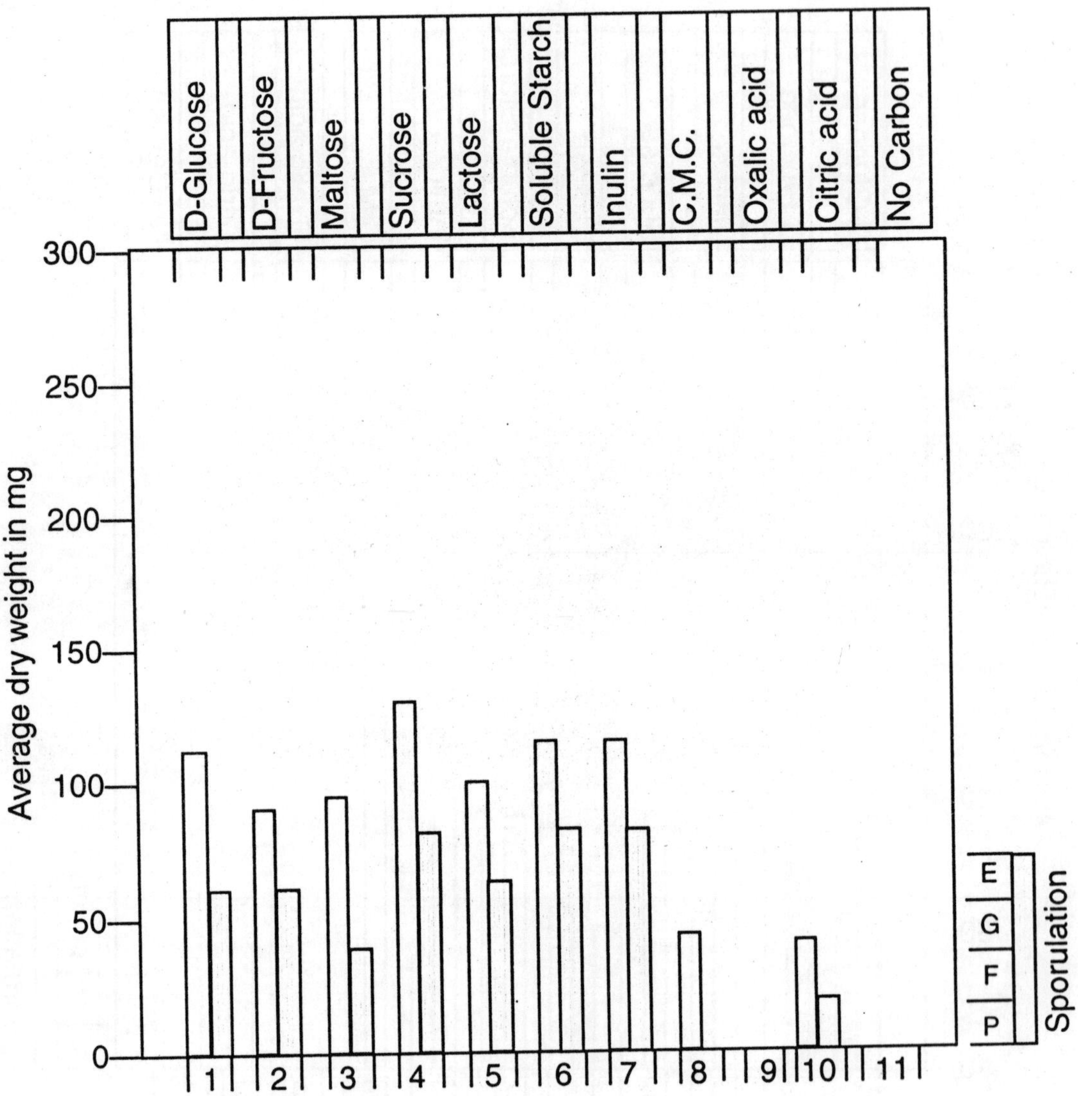

Fig. 10 : Average dry weight and sporulation of *C. cymbiforme* on various carbon sources.

Present study revealed that *C. verruculosus, P. citrinum* and *C. cymbiforme* showed good growth while *A. ustus* showed moderate growth on D-fructose. But sporulation was found excellent in *A. ustus* while good in other three fungi.

According to Cochrane (1958) there appears to be numerous exceptions to the general rule that fructose is equivalent to glucose. Fructose was the poorest hexose for *Phyllosticta morifolia* (Bilgrami and Tandon, 1957). Rai (1982) reported good growth of *Phomopsis* sp. on fructose. Arya (1991) reported poor growth in all three species of *Phomopsis,* whereas, *P. pedilanthi* produced good growth on this keto hexose.

In these hexoses pH drifted towards alkaline side in all the cases.

Maltose, which occurs in nature as a product of starch hydrolysis is reported to be used efficiently by a majority of fungi and even by those which have a limited substrate range. In the present study *C. verruculosus* showed good growth while rest of the fungi showed moderate growth on maltose as a carbon source. Arya (1990-91 & 1991) reported good growth for all the four species of *Phomopsis* similar to Rai (1982) for *Phomopsis sapotae.* Sporulation was found different in all the four fungi. *C. verruculosus* showed poor sporulation while *P. citrinum* showed excellent, *A. ustus* good and *C. cymbiforme* fair on this sugar. Saini and Sidhu (1983) reported best growth of *Choanephora infundibulifera* on maltose.

Sucrose is a common oligosaccharide and a good source of carbon for fungi. Most of the fungi are able to hydrolyse sucrose into glucose and fructose. Thus it is assimilated through hydrolytic pathway. Present investigation revealed that it was the best source of carbon for *C. cymbiforme* like *Phomopsis psidii, P. gulabia* (Arya, 1990-91& 1991) and *P. sapotae* (Rai, 1982). Moderate growth was obtained in the case of *C. verruculosus, P. citrinum* and *A. ustus* like *Phomopsis pedilanthi* (Arya, 1990-91). Sporulation was found poor in *C. verruculosus* while in rest of the fungi sporulation was observed excellent.

Lactose, a mammalian milk sugar, can be hydrolysed into glucose and galactose. Hydrolysis of lactose by fungi with the help of the enzyme in medium is taking more time rather than other disaccharides. That is why lactose is considered as a poor source of carbon. It is evident from the tables 15 to 18 that lactose showed moderate growth in all the four fungi similar to *Fusarium* spp. (Mathew, 1993) and *Phomopsis viticola* and *P. psidii* (Arya, 1991). Sporulation was found good in all the cases except in the case of *C. verruculosus* where it was poor. pH drift was found towards acidic to neutral or slightly alkaline side.

Starch, the universally occurring polysaccharide, is considered to be an excellent source of carbon for most of the fungi (Cochrane, 1958). According to Lilly and Barnett (1951) only those fungi which produce amylase are able to utilize starch; this ability is common but not universal among fungi. In the present study *C. verruculosus* and *C. cymbiforme* induced good growth while *P. citrinum* and *A. ustus* induced moderate growth on this polysaccharide. Mathew (1993) reported starch as a poor source of carbon for all the four species of *Fusarium* except *F. oxysporum* where he found moderate growth, while Arya (1990-91 & 1991) reported good growth of all the four spp. of *Phomopsis*. Rajak (1983) reported excellent growth of *Colletotrichum gloeosporioides* on starch. Sporulation was observed nil in *C. verruculosus* while excellent in rest of the three fungi. At the end of incubation period pH of the media showed a drift towards alkaline side.

Inulin has been reported as a poor source of carbon for most of the fungi. It is probably due to failure of the fungi to secrete enough quantity of enzyme inulase (Thind and Rawla, 1958). In the present investigation it was observed that all the four fungi showed moderate growth on this polysaccharide. Arya (1990-91 & 1991) reported poor growth in all the four spp. of *Phomopsis*. Mathew (1993) reported good growth in *Fusarium udum* and *F. moniliforme*, moderate growth in *F. oxysporum* and poor growth in *F. pallidoroseum* on inulin. Sporulation was observed excellent in *C. cymbiforme*, good in *P. citrinum* and fair in *C. verruculosus* and *A. ustus*. pH drifted towards alkaline side after incubation period.

Carboxymethyl cellulose (CMC) showed good growth. In *C. verruculosus* and *A. ustus* but showed moderate growth in *P. citrinum*, whereas, *C. cymbiforme* showed poor growth on this polysaccharide. Mathew (1993) reported this polysaccharide as a poor source of carbon for all the four spp. of *Fusarium* except *F. oxysporum* which showed moderate growth. Sporulation was nil in all the four fungi. Media remained acidic in all the cases except *C. cymbiforme* which became alkaline after incubation.

Organic acids are generally poor sources of carbon for fungi than the carbohydrates (Cochrane, 1958). In the present study all the four fungi failed to grow on oxalic acid. Sporulation was found nil and pH remained unchanged. Oxalic acid supported moderate growth of *Phomopsis psidii* (Arya, 1991) while *P. sapotae* (Rai, 1982) failed to grow on this acid.

Present study revealed that all the fungi showed moderate growth except *C. cymbiforme* which showed poor growth on citric acid. Sporulation was good in case of *C. verruculosus* and *A. ustus* while poor in case of other two fungi. pH of the media drifted towards alkaline side in all the cases at the end of incubation period. Arya (1991) reported that citric acid was a good carbon source for *Phomopsis viticola, P. pedilanthi* while *P. gulabia* and *P. psidii* produced poor growth like *P. sapotae* (Rai, 1982).

The results clearly substantiate the fact that all the four fungi under present study are specific in their growth and sporulation responses on various carbon sources, and there is no strict correlation between growth and sporulation.

5.5 Utilization of Carbohydrates

The carbohydrates are most common and important source of carbon for fungi. These are chemically defined as aldehyde or ketone derivatives of higher polyhydric alcohols or as compounds which yield these derivatives on hydrolysis. On the basis of sugar units they are classified as mono, oligo and polysaccharides.

Of the monosaccharides glucose is utilized by more fungi than any other sugar. A few species such as *Leptomitus lacteus* is unable to utilize any sugar (Lilly and Barnett, 1951). The oligosaccharides are utilized by some species than is glucose. Most fungi hydrolyse oligosaccharides before utilization. Complex carbohydrates thus split up into monosaccharide units or their derivatives, which subsequently enter into various metabolic pathways. They also take part in the synthesis of reserve food materials of fungal mycelium. Among the polysaccharides cellulose and starch are most abundant in nature. These are hydrolysed or degraded to low molecular weight compounds before utilization.

Chromatographic analysis of the medium was conducted daily upto 15 days to observe the utilization of various carbohydrates. On the basis of this analysis an attempt has been made to establish the pathway of utilization (a)hydrolytic or indirect and (b) non-hydrolytic or direct of di and polysaccharides. The methods used for the study have been given in detail in chapter 3 Investigation techniques and results are summarized in tables 19 to 22 and figs. 11 to 18.

Table-19 : Average dry weight (in mg), final pH and utilization of Monosaccharides and Polysaccharide by Four Fungi.

Sugars Polysaccha-ride	*Days of Incuba-tion*	*C. verruculosus*			*P. citrinum*			*A. ustus*			*C. cymbiforme*		
		Dry weight	*Final pH*	*Pre-sence (days)*	*Dry weight*	*Final pH*	*Pre-sence* (days)	*Dry weight*	*Final pH*	*Pre-sence* (days)	*Dry weight*	*Final pH*	*Pre-sence* (days)
Glucose	5	057	7.0		110	5.5		126	7.0		115	6.0	
	10	117	7.0	12	096	5.5	4	102	7.0	4	128	7.0	15
	15	185	7.0		066	7.0		083	7.0		105	7.5	
Fructose	5	098	7.0		080	7.0		110	7.0		092	7.0	
	10	106	7.0	4	083	7.5	4	072	8.0	4	121	7.5	15
	15	106	7.5		099	7.5	.	077	8.0		097	8.0	
Starch	5	125	5.0		109	5.0		113	6.5		101	6.0	
	10	178	7.0	10	103	6.0	4	104	7.5	4	151	7.0	6
	15	151	7.5		085	6.0		086	7.5		107	7.5	

Table-20 : Average dry weight (in mg), final pH and utilization of Maltose and mixtures of its hydrolytic products by Four Fungi.

Organisms	*Sugars*	*Average dry weight (mg)*			*Final pH*			*Presence of Sugars (days)*	
		5 days	*10 days*	*15 days*	*5 days*	*10 days*	*15 days*	*Maltose*	*Glucose*
C. verruculosus	Maltose	110	120	120	5.5	7.5	7.5	2	3
	*Glucose	057	117	185	7.0	7.0	7.0	-	12
P. citrinum	Maltose	090	065	078	5.5	7.0	7.0	4	10
	*Glucose	110	096	066	5.5	5.5	7.0	-	4
A. ustus	Maltose	070	080	085	5.5	5.5	5.5	5	12
	*Glucose	126	102	083	7.0	7.0	7.0	-	4
C. cymbiforme	Maltose	053	071	079	5.5	6.5	7.0	4	12
	*Glucose	115	128	105	6.0	7.0	7.5	-	15

*As presented in table of mono and polysaccharides (Table 19).

Table - 21 : Average dry weight (in mg), final pH and utilization of Sucrose and mixture of its hydrolytic products by Four Fungi.

Organisms	*Sugars*	*Average dry weight (mg)*			*Final pH*			*Presence of Sugars (days)*		
		5 days	*10 days*	*15 days*	*5 days*	*10 days*	*15 days*	*Suc-rose*	*Glu-cose*	*Fruc-tose*
C. verruculosus	Sucrose	240	230	220	7.0	7.0	7.0	4	6	-
	½ Glucose+ ½ Fructose	043	090	065	7.0	7.0	7.5	-	6	4
P. citrinum	Sucrose	102	095	070	5.0	5.0	7.0	2	4	-
	½ Glucose+ ½ Fructose	077	080	064	7.0	7.0	7.0	-	2	2
A. ustus	Sucrose	115	093	086	6.5	7.0	7.0	2	4	-
	½ Glucose+ ½ Fructose	065	086	055	6.0	7.0	7.5	-	2	2
C. cymbiforme	Sucrose	088	142	128	7.0	7.5	7.5	2	6	15
	½ Glucose+ ½ Fructose	120	100	070	7.0	7.5	-		4	4

Table-22 : Average dry weight (in mg), final pH and utilization of Lactose and mixtures of its hydrolytic products by Four Fungi.

Organisms	*Sugars*	*Average dry weight (mg)*			*Final pH*			*Presence of Sugars (days)*		
		5 days	*10 days*	*15 days*	*5 days*	*10 days*	*15 days*	*Lactose*	*Glucose*	*Galactose*
C. verruculosus	Lactose	000	058	071	6.0	7.0	7.0	15	-	-
	½ Glucose+ ½ Galactose	115	085	094	6.0	6.5	6.5	-	-	4
P. citrinum	Lactose	000	058	078	5.5	6.0	6.0	10	-	-
	½ Glucose+ ½ Galactose	130	084	079	4.5	5.0	5.0	-	-	4
A. ustus	Lactose	086	076	054	7.0	7.5	4	04	-	-
	½ Glucose+ ½ Galactose	133	068	065	7.0	7.5	7.5	-	-	1
C. cymbiforme	Lactose	010	083	097	6.0	7.0	7.0	10	-	-
	½ Glucose+ ½ Galactose	053	071	079	7.0	7.0	7.5	-	1	1

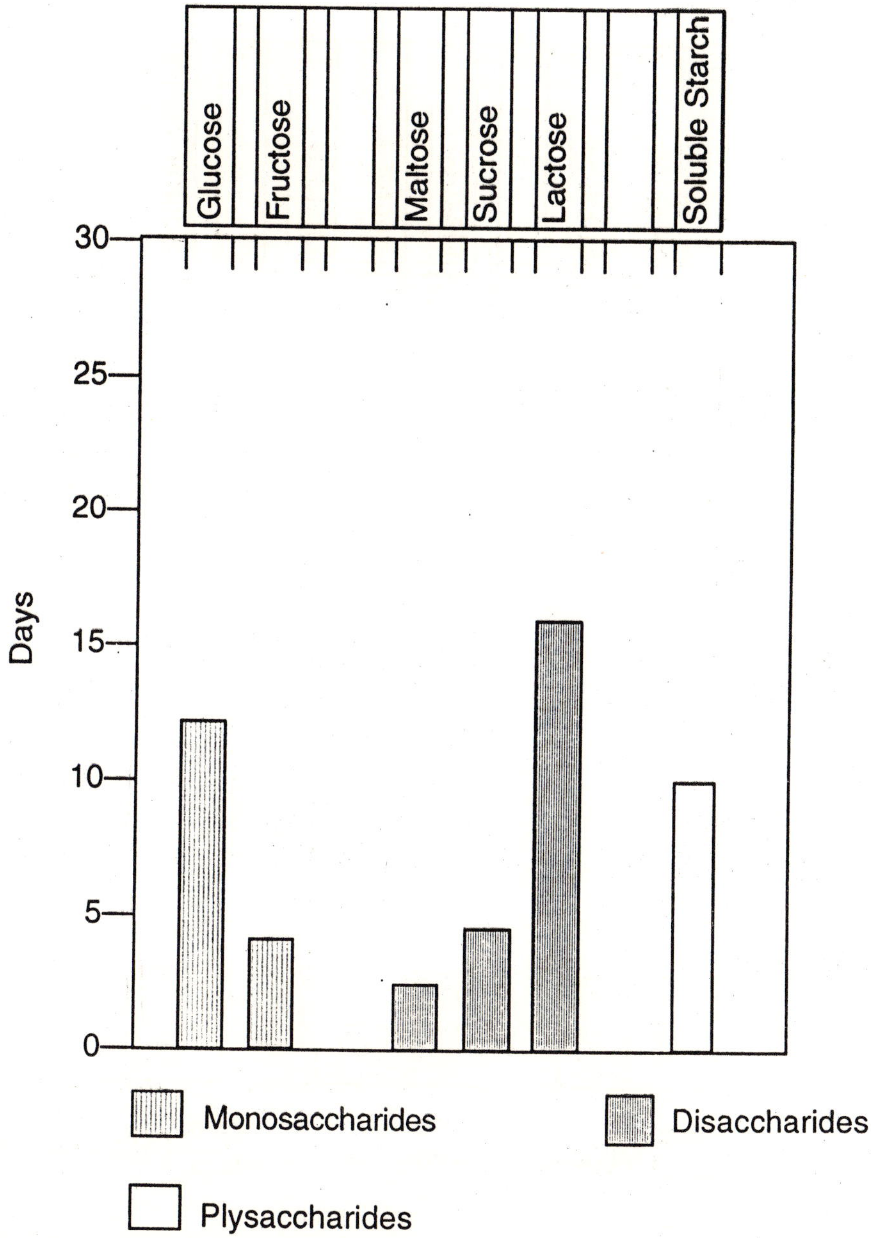

Fig. 11 : Rate of utilisation of Carbohydrates (in days) by *C. verruculosus.*

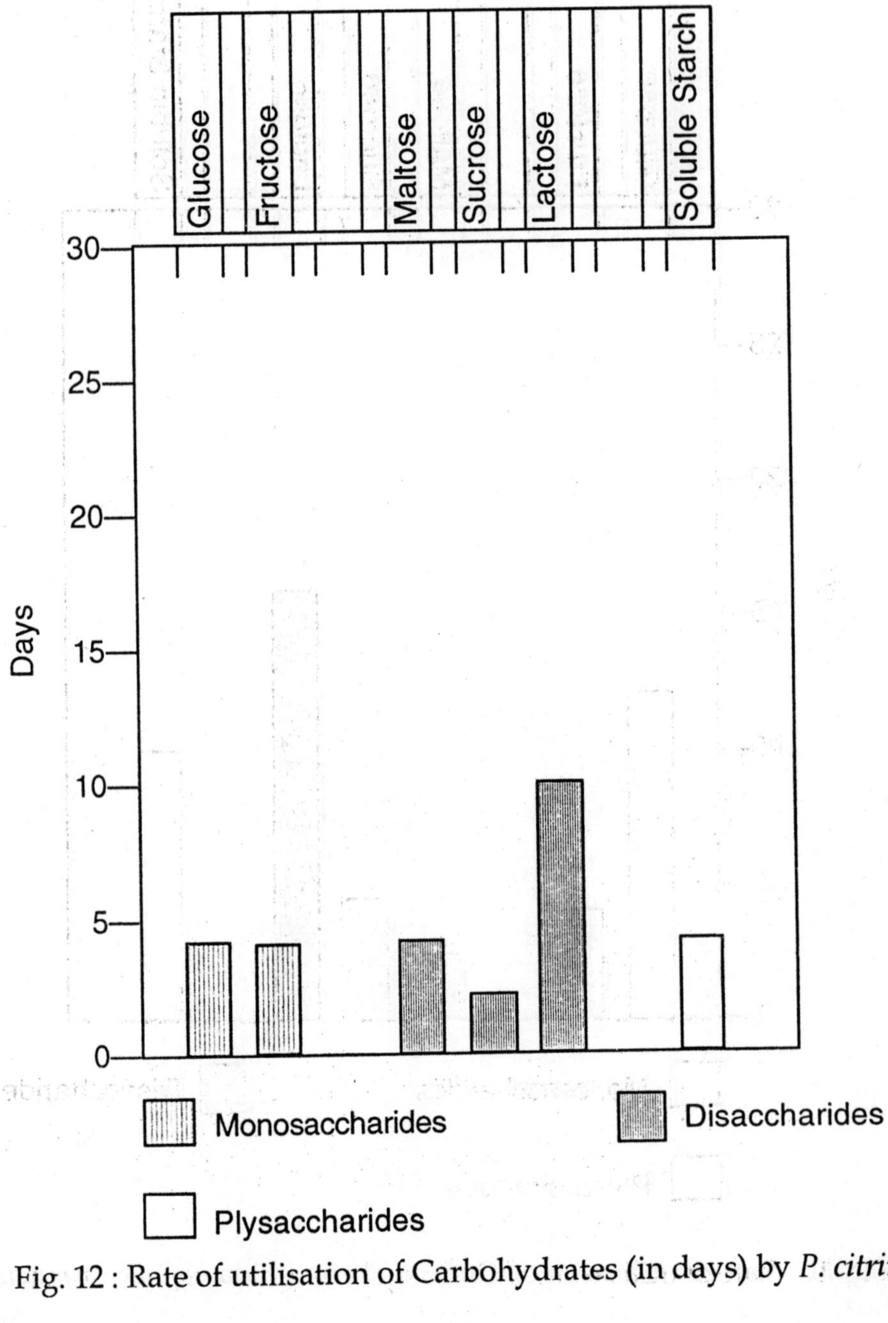

Fig. 12 : Rate of utilisation of Carbohydrates (in days) by *P. citrinum*.

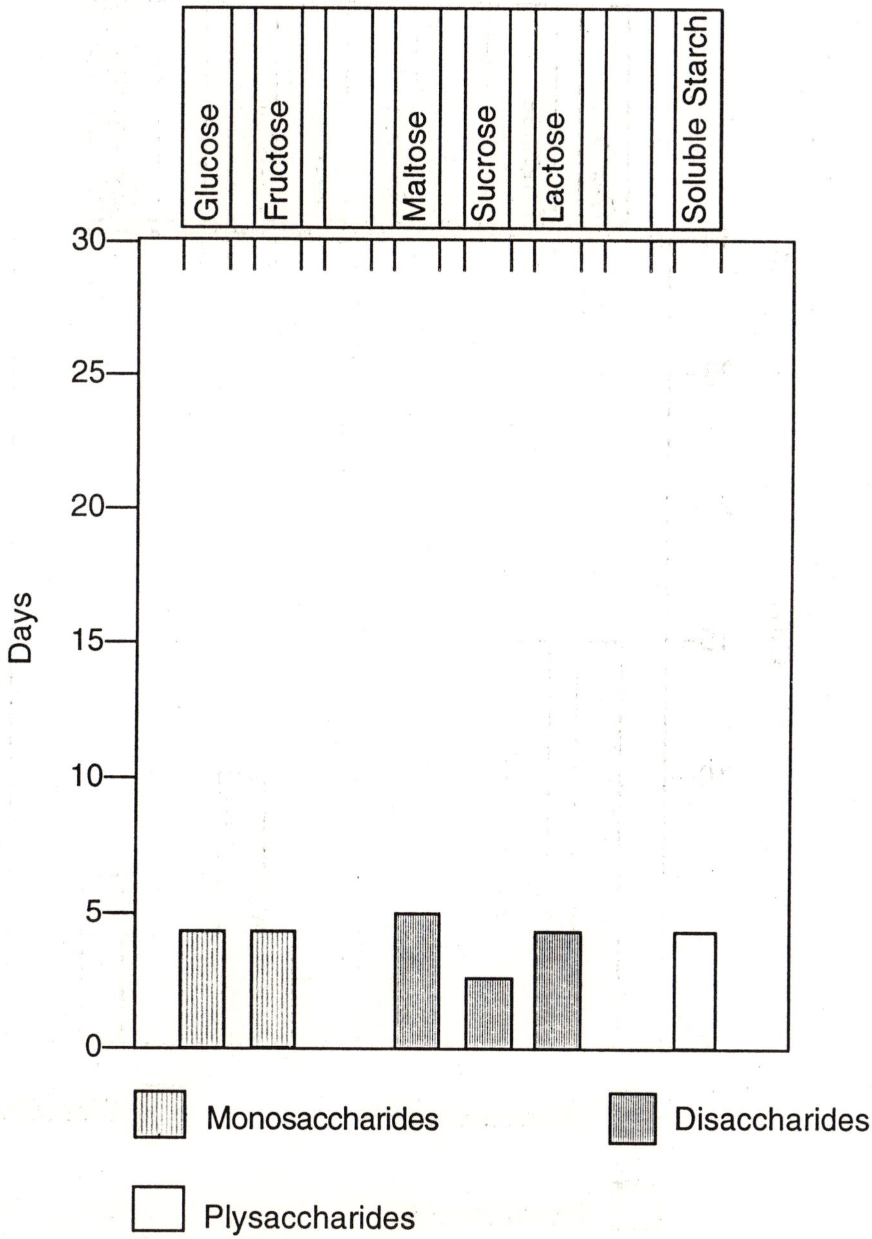

Fig. 13 : Rate of utilisation of Carbodydrates (in days) by *A. ustus*.

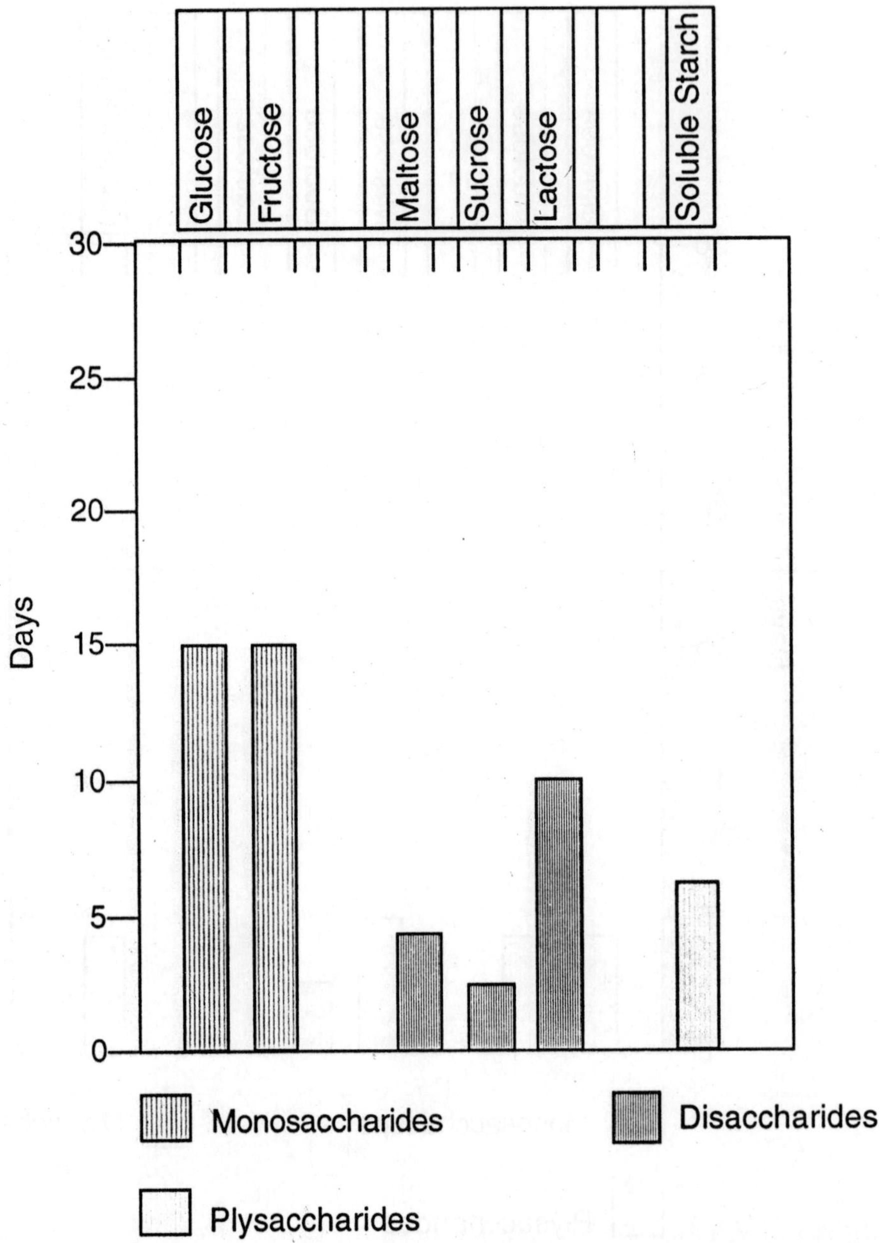

Fig. 14 : Rate of utilisation of Carbohydrtaes (in days) by *C. cymbiforme*.

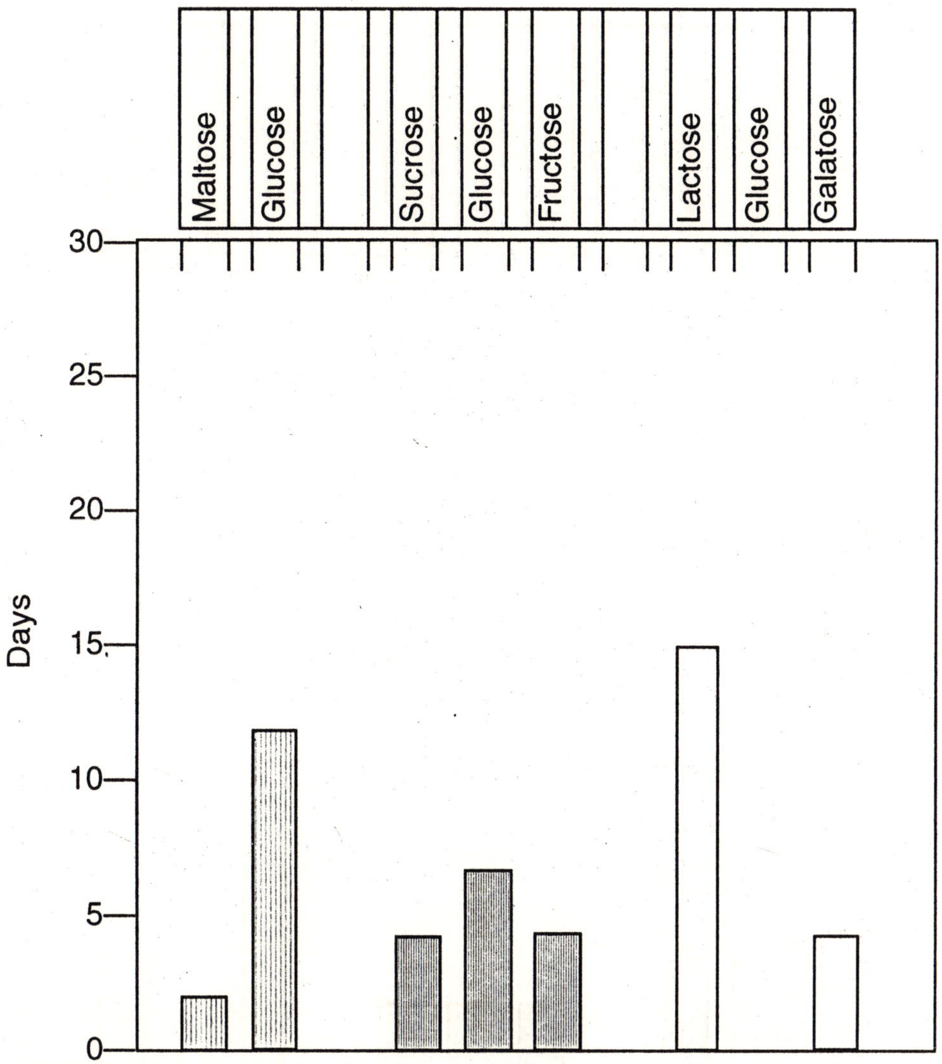

Fig. 15 : Utilisation of Disaccharides and mixture of their hydrolytic products by *C. verruculosus*.

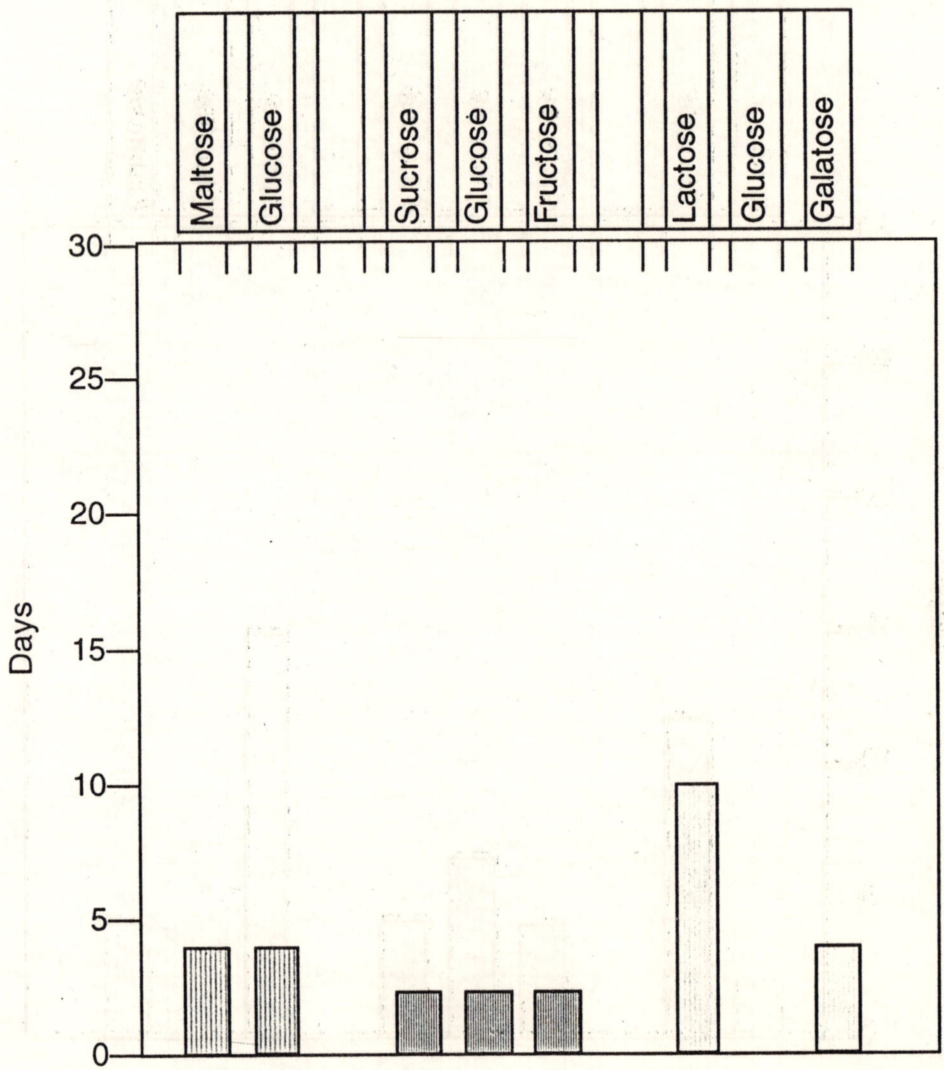

Fig. 16 : Utilisation of Disaccharides and mixture of their hydrolytic products by *P. citrinum.*

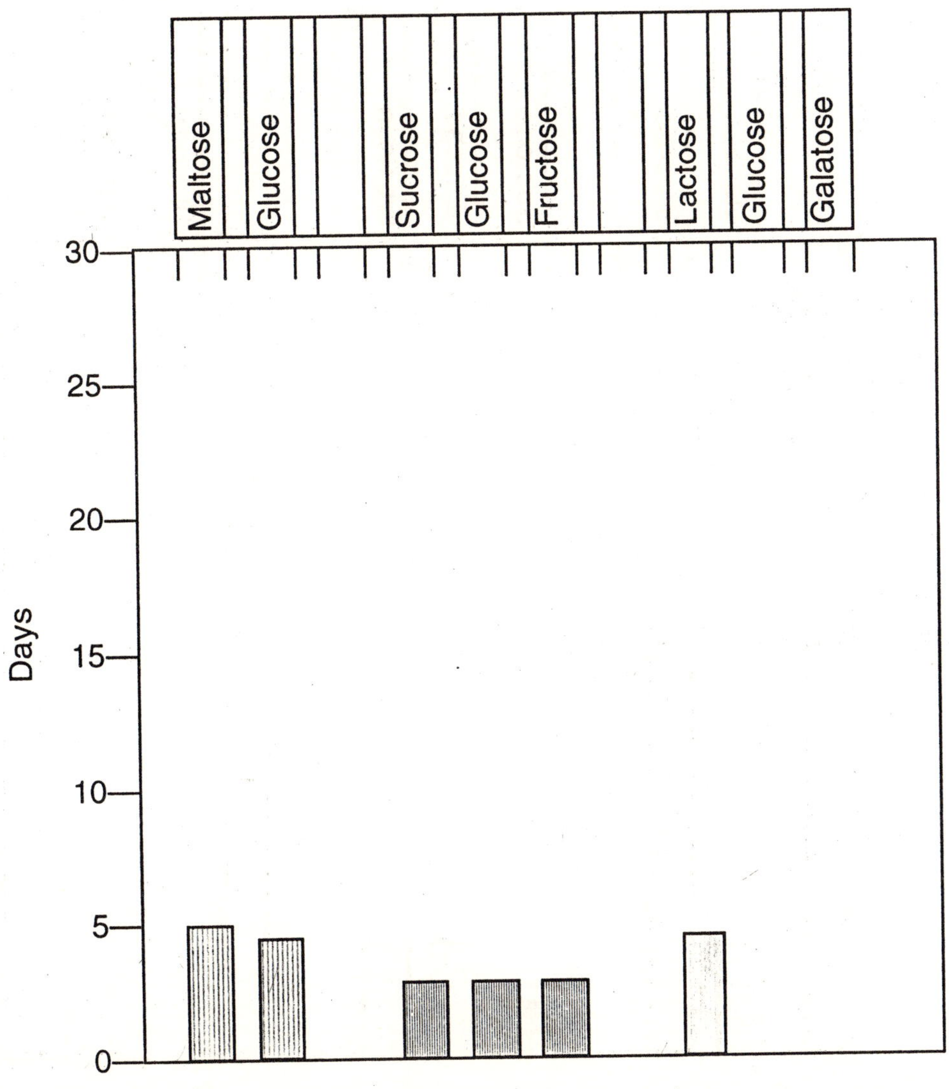

Fig. 17 : Utilisation of Disaccharides and mixture of their hydrolytic products by *A. ustus*.

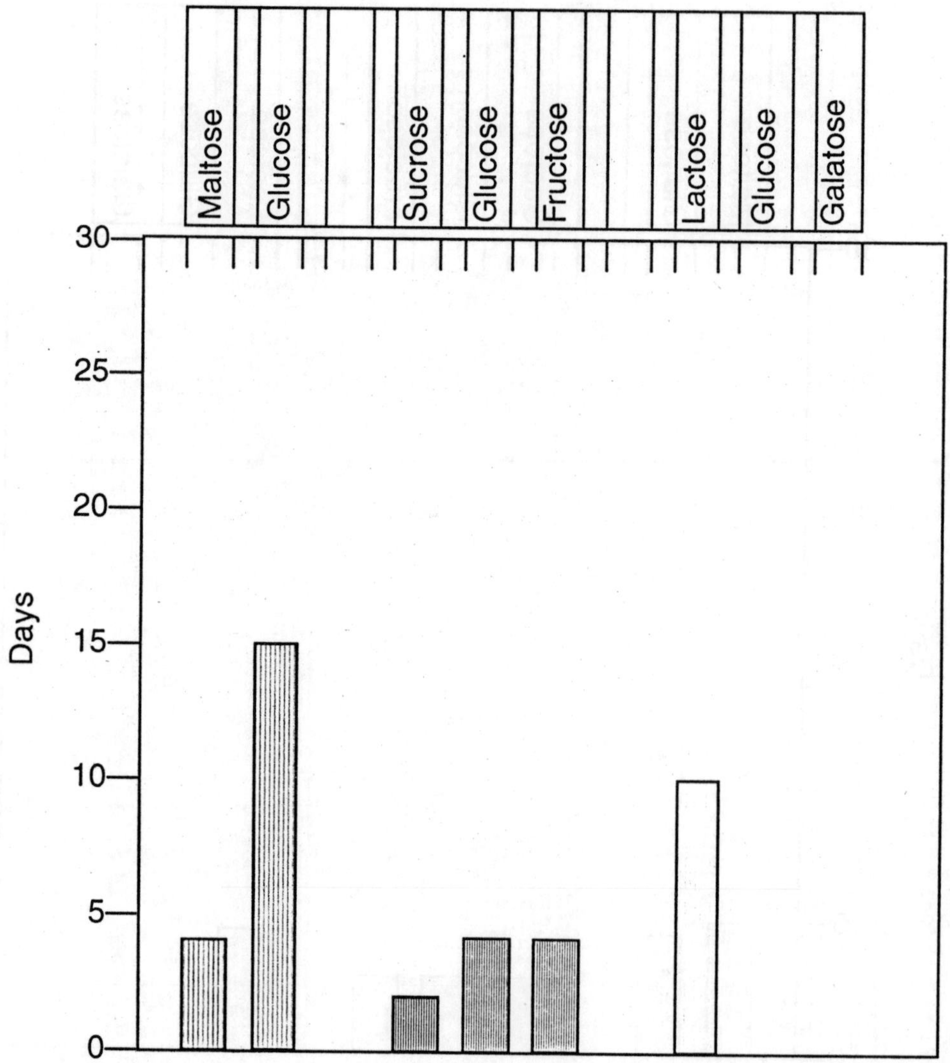

Fig. 18 : Utilisation of Disaccharides and mixture of their hydrolytic products by *C. cymbiforme*.

D-Glucose (Rf 0.60)

Table 19 clearly shows that *P. citrinum* and *A. ustus* utilized this sugar after 4 days, *C. verruculosus* utilized it within 12 days whereas, *C. cymbiforme* could not utilize it within 15 days. Four spp. of *Phomopsis* utilized this sugar variedly (Arya, 1991). *P. viticola* took 9 days while, *P. pedilanthi*, *P. gulabia* and *P. psidii* utilized this sugar in 11, 12 and 13 days respectively.Mycelial weight was highest after 5 days in *P. citrinum* and *A. ustus*, 10 days in *C. cymbiforme* and 15 days in *C. verruculosus*. Final pH of the medium drifted towards netural or alkaline side in all the cases.

D-Fructose (Rf 0.64)

Present study revealed that all fungi except *C. cymbiforme* could utilize it within 4 days. The latter could not utilize it within 15 days of incubation. *P. sapotae* took 7 days for its utilization (Rai, 1982). *P. viticola*, *P. psidii*, *p. gulabia* and *P. pedilanthi* utilized this keto hexose in 5, 7, 9 and 12 days respectively (Arya, 1991).

Sucrose (Rf 0.43)

A large number of workers have shown that most of the fungi are able to hydrolyze disaccharide-sucrose into glucose and fructose and thus it is assimilated through a hydrolytic pathway. However, a few fungi like *Myrothecium verrucaria* (Mandels, 1954) and seven members of Mucorales (Sarbhoy, 1965) were able to consume this sugar through a non-hydrolytic pathway.

In case of *C. verruculosus* sucrose was present up to 4 days, after which it was hydrolysed. Glucose remained present for 6 days. It means that this sugar was hydrolyzed and assimilated within 6 days.

It was observed in case of *C. cymbiforme* that glucose was consumed earlier than fructose similar to *P. viticola* and *P. pedilanthi* (Arya, 1982) and *P. sapotae* (Rai, 1982) while it was reverse in *P. psidii* (Arya, 1982) and *P. vexans* (Chowdhury, 1981).

Maximum dry weight was obtained on 5th day in all the cases except *C. cymbiforme* where it was more on 10th day. pH drifted from acidic to neutral or slightly alkaline side.

Glucose + Fructose (1:1)

Sucrose breaks into two components D-glucose + D-fructose, therefore, it was decided to observe the effect of two sugars in a mixture of 1:1.

Results are indicative that assimilation of component sugars was not similar to sucrose in all the cases. Table 21 clearly depicts that *P. citrinum* and *A. ustus* utilized glucose and fructose within 2 days while *C. verruculosus* took 4 days for fructose and 6 days for glucose. *C. cymbiforme* consumed this mixture within 4 days.

Maximum dry weight was obtained on 10th day except in case of *C. cymbiforme,* where it was maximum on 5th day.

Maltose (Rf 0.40)

Maltose consists of two glucose moiety which are held together by α-1, 4 glucoside linkage. It is utilized by majority of fungi through a hydrolytic pathway. It yields two molecules of glucose when hydrolysis is accomplished by the enzyme glucosidase.

Its presence was detected upto 2 days in *C. verruculosus,* 4 days in *P. citrinum* and *C. cymbiforme* and upto 5 days in *A. ustus.* Its hydrolytic product glucose was detected in the culture filtrates of *C. verruculosus* and *P. citrinum* upto 3 and 10 days respectively while 12 days in *A. ustus* and *C. cymbiforme.* Thus maltose was utilized by the present fungi through a hydrolytic pathway similar to *P. vexans* (Chowdhury, 1981), *P. sapotae* (Rai, 1982) and four spp. of *Phomopsis* (Arya, 1982).

Maximum mycelial weight of *A. ustus* and *C. cymbiforme* was found on 15th day while in *P. citrinum* and *C. verruculosus* it was found maximum on 5th and 10th day respectively.

At the end of incubation period pH of the medium drifted towards neutral to slightly alkaline side in all the cases except in *A. ustus* where it remained acidic.

When the breakdown product of maltose i.e. glucose was provided in the culture medium, it was detected upto 4 days in *P. citrinum* and *A. ustus* and upto 12 days in *C. verruculosus,* while *C. cymbiforme* failed to consume it within 15 days. Maltose and its breakdown products were more preferably utilized by *C. verruculosus* and *C. cymbiforme* while glucose was the choice of other two fungi *P. citrinum* and *A. ustus.*

Lactose (Rf 0.43 B.P.W.)

It is a mammalian sugar. In lactose galactose moity is attached to glucose fraction by a β-linkage. The enzyme lactase, a B-galactosidase has been reported to cause splitting of lactose into glucose and galactose.

It is evident from the table 22 that lactose was present upto 10 days in *P. citrinum* and *C. cymbiforme,* upto 4 days in *A. ustus* and could not be utilized upto 15 days in *C. verruculosus* similar to four fungi investigated by Arya (1982). No breakdown product was detected in any case which is an indication that it was consumed through a non-hydrolytic pathway. The absence of hydrolytic products may be observed due to their simultaneous utilization from the medium.

Maximum dry weight was found on 15th day in all the cases except *A. ustus* where it was maximum on 5th day. Drift in pH was towards neutral or slightly alkaline side.

Glucose + Galactose (1:1)

When the fungi were grown in equimixture of glucose and glactose, glucose was consumed with a faster rate in *C. verrculosus* and *P. citrinum.* Chromatographic studies revealed the presence of galactose in these two fungi upto 4 days. In *A. ustus* and *C. cymbiforme* galactose was present upto one day and glucose was detected in *C. cymbiforme* only upto one day.

Maximum dry weight was obtained on 5th day in all the cases except in *C. cymbiforme* where it was on 15th day. pH drift was from acidic to neutral.

It is interesting to note that except in *C. cymbiforme* all the fungi produced more growth in a mixture of two component sugars of lactose.

Starch

Starch consists of glucose residues joined through glycoside linkages and thus may be thought consisting as repeating units of maltose. Starch consists of two types of molecules. The linear portion of starch is called amylose, while the branched chain fraction is known as amylopectin.

The enzymatic hydrolysis of starch may be represented schematically as follows:

Starch ------> Dextrin -------> Maltose -------> D-glucose

Starch was present upto 4 days in *P. citrinum* and *A. ustus,* 6 days in *C. cymbiforme* and 10 days in *C. verruculosus.* Certain other fungi like *P. viticola* and *P. psidii* also hydrolysed the starch in 10 days (Arya, 1982).

An increase in dry weight of *P. citrinum* and *A. ustus* was recorded upto 5 days. It increased upto 10 days in *C. verruculosus* and *C. cymbiforme.*

The drift in pH was found towards alkaline side.

5.6 Effect of Different Nitrogen Sources

Nitrogen is an essential element for fungi. It has structural as well as functional importance and occurs in fungi in the form of complex organic nitrogenous compounds. Both organic and inorganic forms of nitrogen are available to fungi from nature but as far as utilization is concerned they intrinsically contradict from each other in their metabolic potentialities.

Fungi require nitrogen for the synthesis of a variety of critically important cellular constituents. Most widely used sources of nitrogen are nitrates, nitrites, ammonia, urea and amino acids; although fungi can obtain nitrogen from a wide variety of compounds. Most of the compounds are broken down, at least in parts, to ammonium before being assimilated. Effective utilization of amino acids by fungi depends on the amount and type of carbon source because amino acids serve as a source of carbon as well as a source of nitrogen also (Pathak and Dhawan 1989).

A nitrogen nutrition may affect the carbon uptake. Park (1976) investigated a cellulolytic fungus from an aquatic environment and observed that it decomposed cellulose faster when the mineral nutrients were diluted. Nyuksha (1994) studied nutritive value of some carbon and nitrogen compounds for *Myxotrichum deflexum*. He observed that Asparagine, potassium nitrite and ammonium compounds inhibited development and colouration.

According to Lilly and Barnett (1951) not all nitrogen sources are equally suitable for all fungi. Fungi may be specific in the nitrogen sources they utilize. Nitrates are also used as a nitrogen source by some fungi while they are toxic to many species. Urea is a very good source of nitrogen, which involves intra cellular conversion to ammonium. Urea supports good growth, good conidial development and moderate sclerotial development (Agnihotri, 1968).

There are some other sources of nitrogen which include proteins, peptides, amides, amino acids, purines, pyrimidines, amines and thiourea. The ability to utilize proteins and shorter polypeptides depend on the secretion of proteases that hydrolyze these macromolecules in smaller sub units (Cohen, 1973).

"No single pattern of nitrogen assimilation can be described to apply for all fungi" (Cochrane, 1958). According to previous reports, the amino acids are converted into ammonia before their assimilation by fungi, if amino acids are used as a nitrogen source. Lilly and Barnett (1951) stated that nitrogen is liberated in the form of ammonia during the process of deamination and it is then utilized by most fungi.

Robbins (1937) has classified fungi according to their ability to utilize different sources of nitrogen:

(1) Fungi are able to utilize atmospheric nitrogen, nitrite nitrogen, ammonium nitrogen and organic nitrogen;

(2) Fungi are able to utilize nitrate nitrogen; ammonium nitrogen and organic nitrogen but are not able to utilize atmospheric nitrogen;

(3) Fungi are able to utilize ammonium and organic nitrogen and are unable to utilize atmospheric or nitrate nitrogen;

(4) fungi which are able to utilize only organic nitrogen and unable to utilize atmospheric, nitrate or ammonium nitrogen.

Above classification is very useful in preparing media and in discovering the causes and failure of some fungi to grow on certain media.

In present study the effect of various nitrogen compounds on growth and sporulation of the four fungi has been investigated by taking dry weight of the individual organism as growth parameter. Following nitrogen compounds were used for the present study.

1. Potassium Nitrate, KNO_3
2. Magnesium Nitrate, $Mg(NO_3)_2 . 2H_2O$
3. Ammonium Nitrate, $NH_4 NO_{3\ 2} . 6H_2O$
4. Cobalt Nitrate, $Co(NO_3)_2 . 6H_2O$
5. Copper Nitrate, $Cu(NO_3)_2 . 3H_2O$
6. Potassium Nitrite, KNO_2
7. Urea, $CO(NH_2)_2$
8. Thio urea, $(NH_4)_2CS$
9. Acetamide. $CH_3 CONH_2$

A set of medium was also prepared without any source of nitrogen as a control to observe the growth and sporulation of the organisms under investigation. Results are recorded in table 23 to 26 and figs. 19 to 22.

Table - 23 : Average dry weight (in mg), sporulation and change in pH by *C. verruculosus* on various Nitrogen sources.

Sl. No.	*Nitrogen sources*	*Dry Weight*	*Sporu-lation*	*Final pH*
1	Potassium Nitrate	185	Good	7.0
2	Magnesium Nitrate	085	Poor	7.0
3	Ammonium Nitrate	065	Nil	7.0
4	Cobalt Nitrate	000	Nil	5.5
5	Copper Nitrate	000	Nil	5.5
6	Potassium Nitrite	056	Fair	7.5
7	Urea	063	Fair	7.5
8	Thio Urea	000	Nil	6.0 Spores found dead
9	Acetamide	072	Poor	5.5
10	No nitrogen	020	poor	7.0
	G.M.	52.60		

Summary of dry weight results and conclusion at 5% level of P.

Treatments	:	Highly significant
Replicates	:	Non-significant
S.E.	:	6.1788
C.D. at 5% level of P.	:	$\pm$ 12.8890
Dry weight results	:	$1 > 2 > \overline{9\ 3\ 7}$ $\underline{}$ $> 10 > \overline{4\ 5\ 8}$

Table - 24 : Average dry weight (in mg), sporulation and final pH of *P. citrinum* on various Nitrogen sources.

Sl. No.	*Nitrogen sources*	*Dry Weight*	*Sporu-lation*	*Final pH*
1	Potassium Nitrate	99	Excellent	7.0
2	Magnesium Nitrate	52	Good	6.0
3	Ammonium Nitrate	61	Excellent	4.0
4	Cobalt Nitrate	00	Nil	5.5
5	Copper Nitrate	00	Nil	5.5
6	Potassium Nitrite	42	Good	7.5
7	Urea	56	Good	7.5
8	Thio Urea	00	Nil	6.0
				Spores found alive
9	Acetamide	72	Excellent	6.5
10	No Nitrogen	37	Poor	5.0
	G.M.	41.90		

Summary of dry weight results and conclusion at 5% level of P.

Treatments	:	Highly significant
Replicates	:	Non-significant
S.E.	:	2.8713
C.D. at 5% level of P.	:	$\pm$ 5.9895
Dry weight results	:	$1 > 9 > \overline{3\ \underline{7\ 2}} > \overline{6\ 10} > \overline{4\ 5\ 8}$

Table - 25 : Average dry weight (in mg), sporulation and final pH of *A. ustus* on various Nitrogen sources.

Sl. No.	*Nitrogen sources*	*Dry Weight*	*Sporu-lation*	*Final pH*
1	Potassium Nitrate	90	Excellent	7.0
2	Magnesium Nitrate	69	Good	6.0
3	Ammonium Nitrate	78	Good	7.0
4	Cobalt Nitrate	00	Nil	5.5
5	Copper nitrate	00	Nil	5.5
6	Potassium Nitrite	48	Poor	7.5
7	Urea	42	Good	7.5
8	Thio Urea	00	Nil	6.0 Spores found alive
9	Acetamide	52	Poor	7.0
10	No Nitrogen	25	Poor	5.0
	G.M.	42.20		

Summary of dry weight results and conclusion at 5% level of P.

Treatments	:	Highly significant
Replicates	:	Non-significant
S.E.	:	3.4832
C.D. at 5% level of P.	:	$\pm$ 7.2661
Dry weight results	:	$1 > 3 > 2 > \overline{9\,\underline{6}}\,\underline{7} > 10 > \overline{4\,5\,8}$

Table - 26 : Average dry weight (in mg), sporulation and final pH of *C. cymbiforme* on various Nitrogen sources.

Sl. No.	*Nitrogen sources*	*Dry Weight*	*Sporu- lation*	*Final pH*
1	Potassium Nitrate	110	Good	7.0
2	Magnesium Nitrate	020	Fair	6.0
3	Ammonium Nitrate	072	Excellent	7.0
4	Cobalt Nitrate	000	Nil	5.5
5	Copper Nitrate	000	Nil	5.5
6	Potassium Nitrite	074	Good	7.5
7	Urea	070	Good	7.5
8	Thio Urea	000	Nil	6.0
				Spores found alive
9	Acetamide	083	Good	7.5
10	No Nitrogen	035	Poor	5.0
	G.M.	26.40		

Summary of dry weight results and conclusion at 5% level of P.

Treatments	:	Highly significant
Replicates	:	Non-significant
S.E.	:	2.2558
C.D. at 5% level of P.	:	±4.7057
Dry weight results	:	$1 > 9 > \overline{6\ 3\ 7} > 10 > 2 > \overline{4\ 5\ 8}$

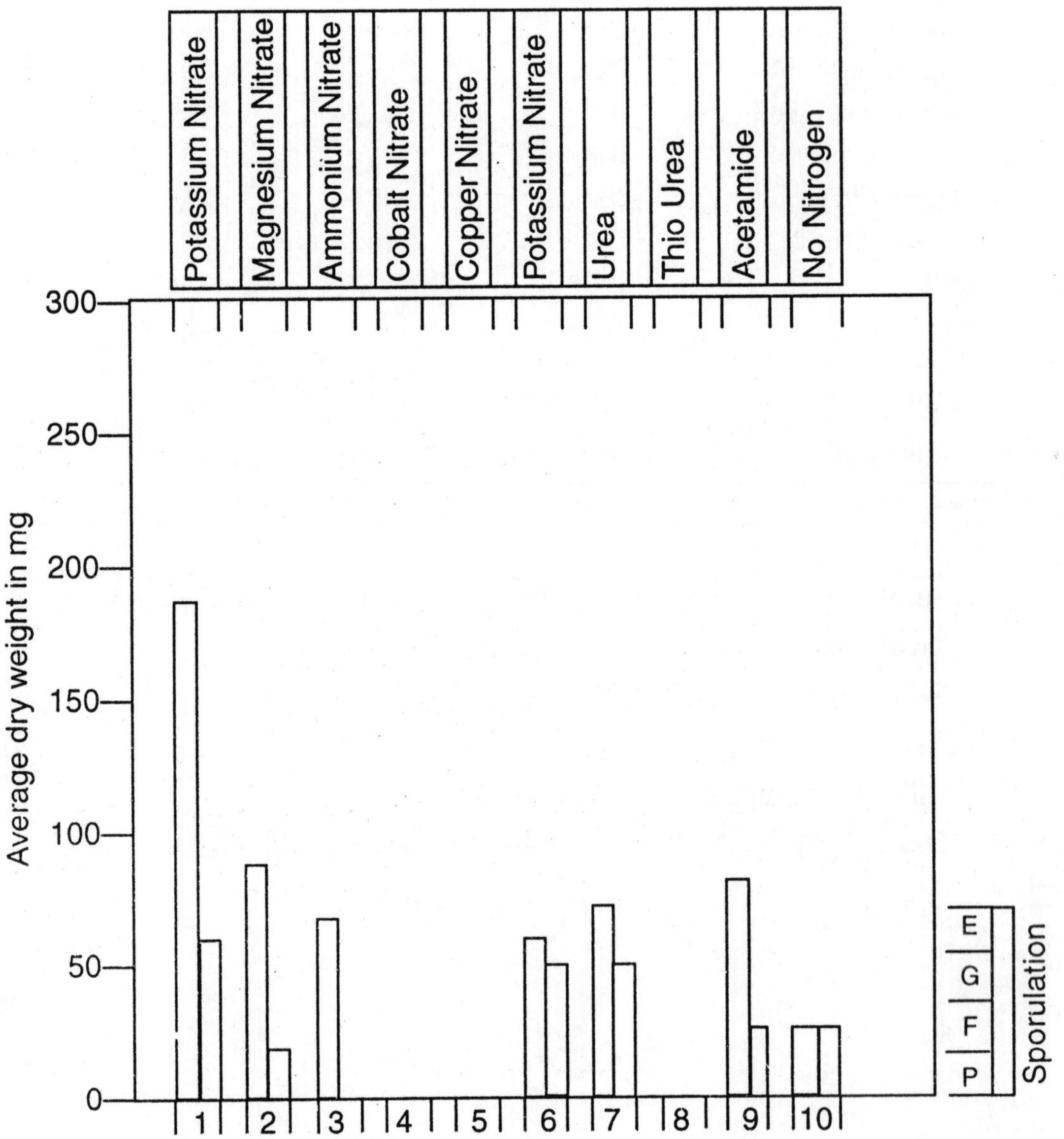

Fig. 19 : Average dry weight and sporulation of *C. verruculosus* on various Nitrogen sources.

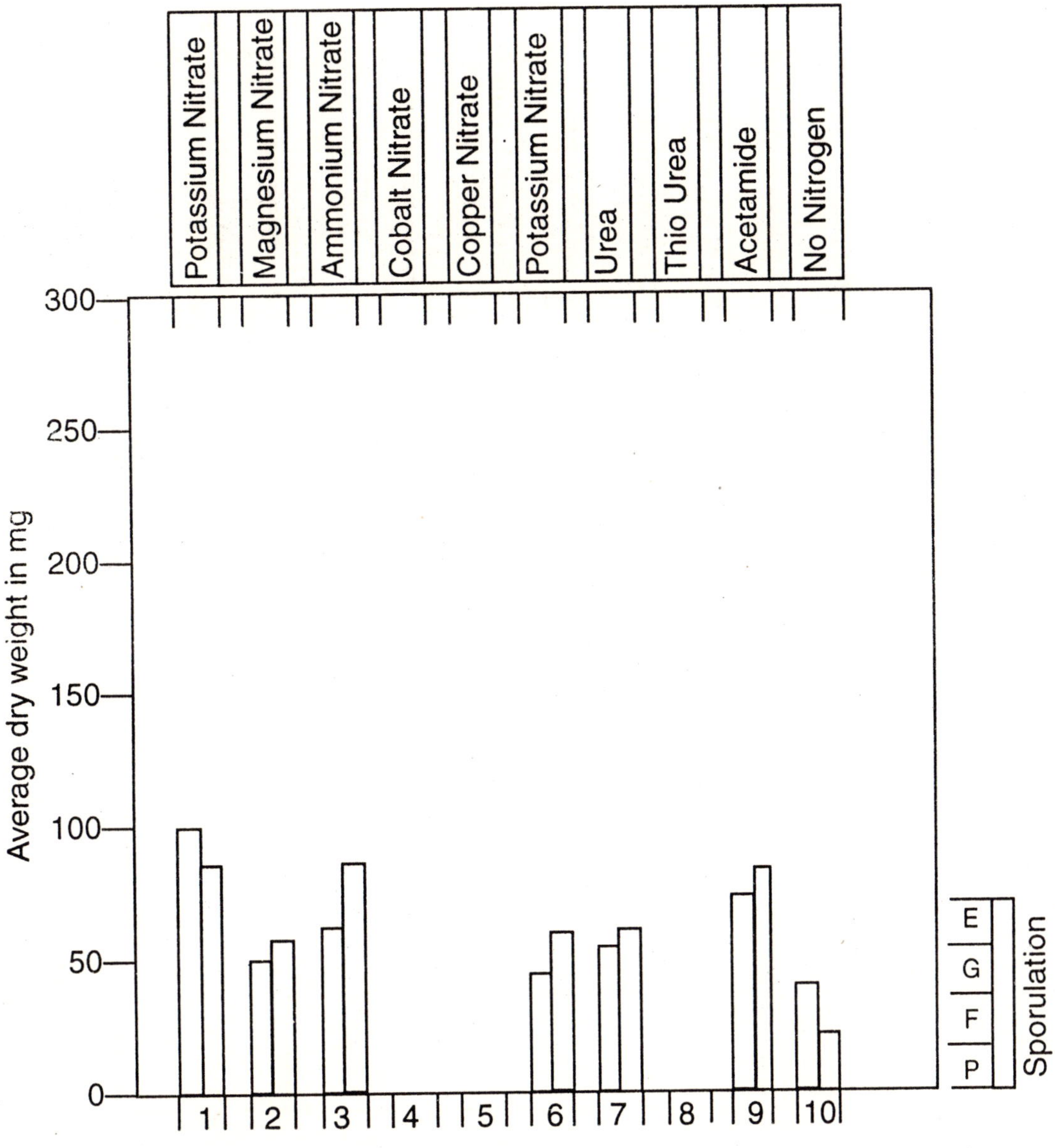

Fig. 20 : Average dry weight and sporulation of *P. citrinum* on various Nitrogen sources.

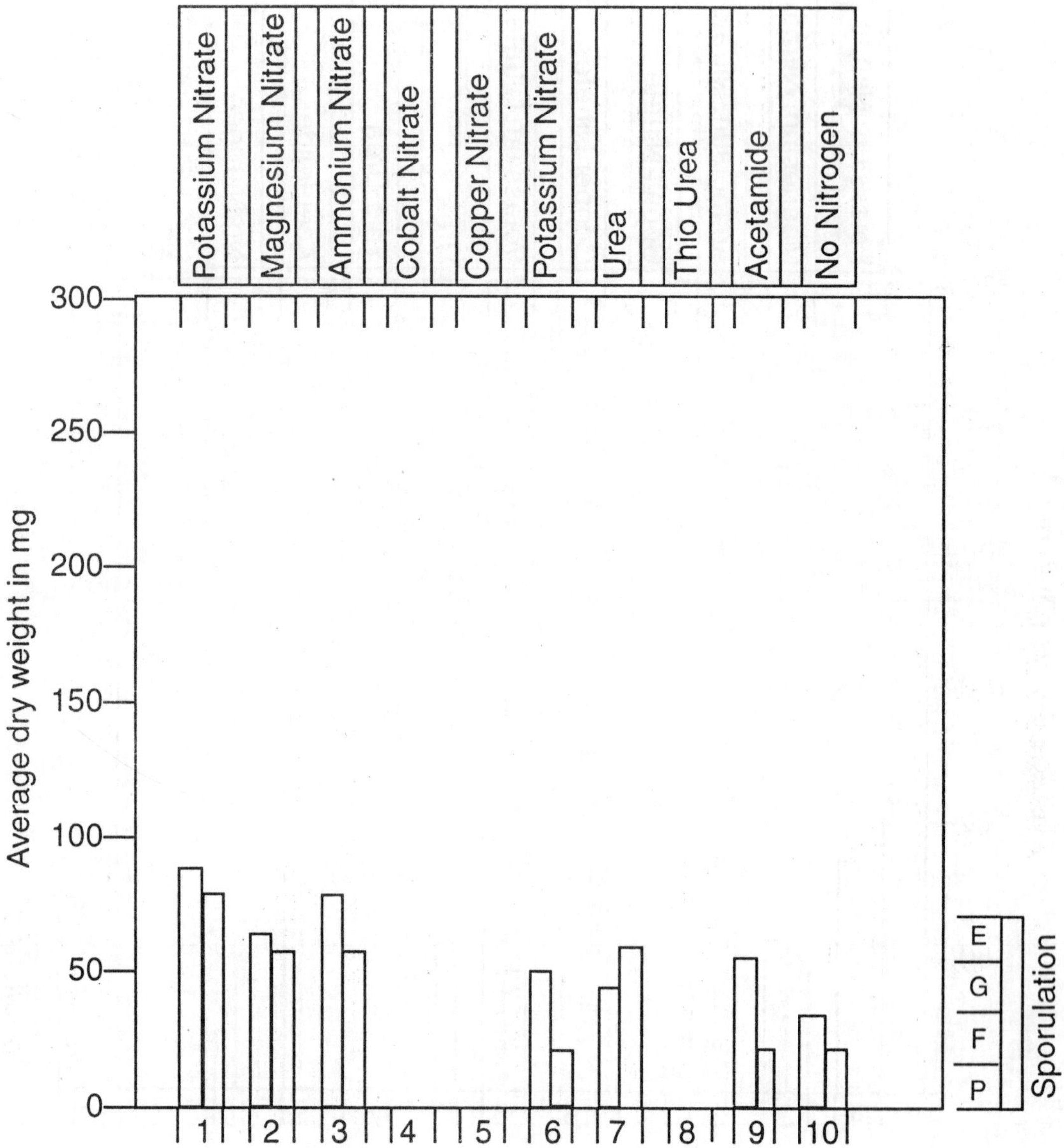

Fig. 21 : Average dry weight and sporulation of *A. ustus* on various Nitrogen sources.

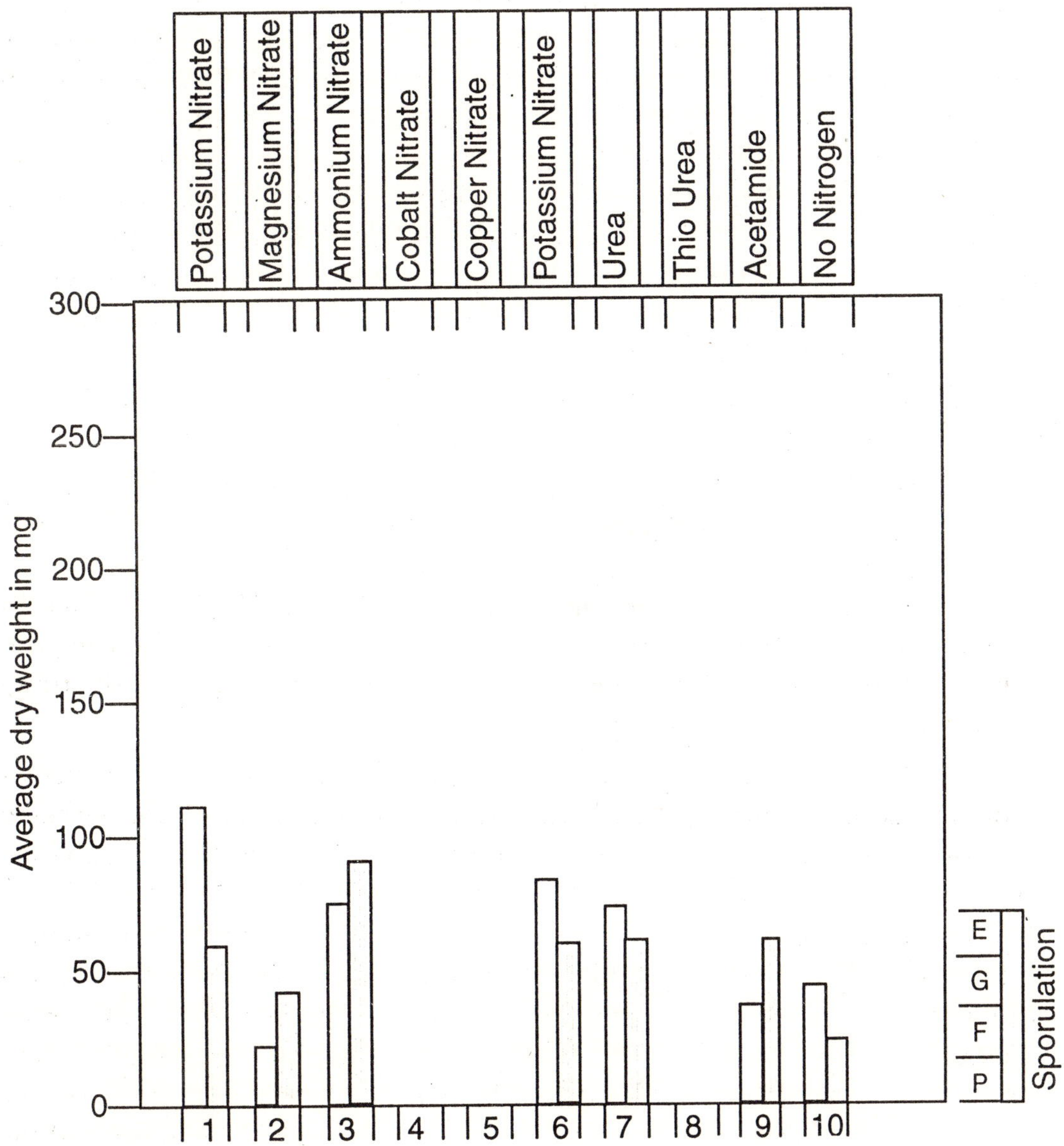

Fig. 22 : Average dry weight and sporulation of *C. cymbiforme* on various Nitrogen sources.

Nitrates occur in the soil and thus are a natural source of nitrogen. A fungus which utilizes nitrate nitrogen must be able to reduce the nitrogen to the oxidation level of ammonia. We may assume that failure of a fungus to utilize nitrate nitrogen is coupled with inability to perform this reduction. During the experiment good growth was observed on nitrates of magnesium, ammonium and potassium whereas, fungi failed to grow on nitrates of cobalt and copper. It is evident from the results that nitrate nitrogen of potassium proved the best for growth in all the four fungi. Lilly and Barnett (1951) reported that *Aspergillus* spp., *Chaetomium cochilioides*, *C. convolutum*, *C. globosum*, *Penicillium* spp. were able to utilize nitrate nitrogen. Gumargalieve *et al.* (1991) have experimented and reported that sodium nitrate increased the growth of *Phytophthora infestans*. Umalkar (1980) studied the influence of different nitrogen sources on *Pestalotia annonicola* and *P. gracilis*. He reported that KNO_3, $NaNO_3$and Ca $(NO_3)_2$ were utilized by both the fungi but nitrites of some elements were toxic to the same fungi.

Owing to the instability of nitrites in acid solution and the destructive effect of nitrous acid on proteins and amino acids, nitrite nitrogen is rarely used in making media. Nitrite is produced by many fungi from nitrate and may accumulate in the medium under certain conditions. The toxic effect is related to the pH of the medium, being greatest at low pH. During the experiment nitrite nitrogen of potassium showed moderate growth in all the fungi under study.

In the present investigation urea, thiourea and acetamide were tested as an organic source of nitrogen. Experiment showed that thiourea was lethal to *C. verruculosus* as in it spores were found dead after 15 days of incubation period, while other organisms showed no growth but their spores were found alive after incubation period. Urea and acetamide showed moderate growth in all the organisms. Adisa (1988) reported that all the organic nitrogen sources supported good growth of *C. verruculosa*. Rajak (1983) reported KNO_3 and asparagine as the best sources of nitrogen for *Colletotrichum dematium* and *C. gloeosporioides* respectively. L-asparagine was also found as a good source of nitrogen for *Fusarium udum* (Mathew, 1993). Ansari *et al.* (1989) studied the effect of various nitrogen sources on growth and sporulation for *Alternaria brassicae*. They reported maximum growth on peptone and minimum on ammonium molybdate.

Experiment without any source of nitrogen showed negligible growth in all the four organisms which proved that nitrogen is essential element for the life of the organism.

The nutritional conditions under which fungi sporulate are quite different from those which are optimum for vegetative growth. In the present investigation all the nitrogen sources are not equally suitable for sporulation. Some, which are favourable for mycelial growth do not favour sporulation. In *C. verruculosus* mycelial growth was found more in magnesium and ammonium nitrates as well as in acetamide but sporulation ranged from poor to nil. In *P. citrinum* sporulation was found good to excellent, while in other two organisms sporulation was found good to nil. Sporulation was good to excellent only in the case of potassium nitrate in all the organisms.

In some compounds a drift in pH was observed towards acidic side while in rest it drifted towards neutral or alkaline side.

5.7 Effect of Different Sulphur Sources

Sulphur, an important constituent of protein, plays a significant role in the metabolic activities of fungi. It is an essential element for the biosynthesis of prosthetic (-SH) group containing enzymes, co-enzymes and vitamins which are indespensable for many vital processes of the fungi (Lilly and Barnett, 1951; Cochrane, 1958; Tandon, 1961; Bhargava and Tandon, 1963; Arya, 1982).

According to Pathak and Dhawan (1989) sulphur is required as macronutrient because of its role as a component of protein and vitamins. Several workers have demonstrated that sulphate is most common inorganic form of sulphur utilized by various fungi like *Penicillium chrysogenum* (Yamamoto and Segal, 1966) and *Phomopsis viticola* (Arya, 1991). Fungi vary considerably in their ability to utilize different sources of sulphur.

Sulphur is present in many types of compounds both inorganic and organic. The state of oxidation of sulphur as well as the specific structure of organic sulphur compounds affects utilization. Sulphate sulphur $-SO_4$ is the most common source of sulphur used in media. Some fungi, however, require specific organic sources of sulphur.

In comparision with general nutritional behaviour, fungi show variations in their choice and ability to utilize different sources of sulphur (Hawker, 1950, Lilly and Barnett, 1951). Sulphides, thiosulphates, thiourea and a variety of sulphur sources are easily utilized by many fungi. Some fungi can even utilize elemental sulphur (Devey and Papavizas, 1962). Cochrane (1958) stated that most fungi can supply all their needs of sulphur from inorganic sulphate i.e. they reduce the sulphate and incorporate it into organic molecules.

Steinberg (1941) has made a comprehensive study of sulphur requirement of *Aspergillus niger* and has concluded that oxidized sulphur containing inorganic compounds were utilized. While sulphide and disulphides were unutilizable.

Fisher distinguished two groups of micro-organisms viz. Euthiotrophic and Parathiotrophic. Out of these the organisms belonging to the first group have the ability to utilize sulphate sulphur, while the organisms of the second group have the ability to utilize reduced sulphur only.

Many of the fungi appear to be more versatile as far as their ability to utilize sulphur source is concerned. For instance, *Pencillium chrysogynum* (Hockenhull, 1948), and *Pestalotia malorum* (Tandon, 1950) are some of the fungi which can utilize a variety of sulphur sources.

Our knowledge about the relative importance of different sulphur sources is still very inadequate. It was, therefore, considered desirable to study the influence of different sources of sulphur on the growth and sporulation of fungi under present study. The following sulphur sources were used for this study.

Copper Sulphate	$CuSO_4 . 5H_2O$
Ferrous Sulphate	$FeSO_4 . 7H_2O$
Potassium Persulphate	$K_2S_2O_8$
Potassium Metabisulphite	$K_2S_2O_5$
Thiourea	$(NH_2)_2 CS$
Ammonium Sulphate	$(NH_4)_2 SO_4$
Magnesium Sulphate	$MgSO_4 . 7H_2O$
Sodium Sulphate (Anhy.)	Na_2SO_4
Sodium Metabisulphite	$Na_2S_2O_5$
Sodium Thiosulphate	$Na_2S_2O_3 . 5H_2O$

A medium was also prepared without adding any source of sulphur, so as to findout the effect of complete absence of sulphur on different fungi under study. The results are summarized in tables 27 to 30 and figs. 23 to 26.

The results revealed that all the sulphur sources were not equally suitable for the growth and sporulation of the four organisms under present study. Magnesium sulphate supported good growth and sporulation in all the organisms except *A. ustus* similar to *Chaetosphaeropsis* sp. (Rai, 1982). Many workers like Lilly and Barnett (1951), Tandon and Bilgrami (1958) and Hasija (1969) reported that for most of the fungi, sulphate sulphur appears to be a most favourable source. Compared to magnesium sulphate and other sulphate salts such as ferrous sulphate, potassium persulphate and sodium sulphate showed moderate growth in all the organisms. Potassium and sodium bisulphite showed nearly equal growth in all the organisms. Thiourea and copper sulphate showed a negligible growth, therefore, they are toxic to the organisms. Without any sulphur source negligible growth appeared in the flasks.

Table - 27 : Average dry weight (in mg), sporulation and final pH of *Cochliobolus verruculosus* on different sulphur sources.

Sl. No.	*Nitrogen sources*	*Dry Weight*	*Sporulation*	*Final pH*
1	Copper Sulphate	024	Poor	5.5
2	Ferrous Sulphate	115	Poor	7.5
3	Potassium Persulphate	070	Fair	7.5
4	Potassium Metabisulphite	060	Poor	7.5
5	Thiourea	044	Poor	4.0
6	Ammonium Sulphate	060	Poor	6.5
7	Magnesium Sulphate	185	Good	7.0
8	Sodium Sulphate (Anhy.)	080	Fair	9.0
9	Sodium Metabisulphite	052	Fair	7.0
10	Sodium thiosulphate	075	Good	7.0
11	No Sulphur	035	Nil	6.5
	G.M.	72.72		

Summary of dry weight results and conclusion at 5% level of P.

Treatments	:	Highly significant
Replicates	:	Non-significant
S.E.	:	4.6839
C.D. at 5% level of P.	:	± 9.7145
Dry weight results	:	7 > 2 > 8 10 3 4 6 9 5 11 > 1

Table - 28 : Average dry weight (in mg), sporulation and final pH of *Penicillium citrinum* on different sulphur sources.

Sl. No.	*Nitrogen sources*	*Dry Weight*	*Sporu-lation*	*Final pH*
1	Copper Sulphate	000	Nil	5.0
2	Ferrous Sulphate	092	Nil	5.5
3	Potassium Persulphate	069	Poor	4.5
4	Potassium Metabisulphite	044	Poor	6.0
5	Thiourea	031	Poor	4.5
6	Ammonium Sulphate	044	Fair	6.5
7	Magnesium Sulphate	099	Good	7.0
8	Sodium Sulphate (Anhy.)	062	Poor	5.5
9	Sodium Metabisulphite	044	Fair	7.5
10	Sodium Thiosulphate	055	Fair	6.5
11	No Sulphur	050	Nil	4.5
	G.M.	53.63		

Summary of dry weight results and conclusion at 5% level of P.

Treatments	:	Highly significant
Replicates	:	Non-significant
S.E.	:	2.8390
C.D. at 5% level of P.	:	$\pm$ 5.8881
Dry weight results	:	7 > 2 > 3 > 8 > 10 > 11 > $\overline{4\ 6\ 9}$ > 5 > 1

Table - 29 : Average dry weight (in mg), sporulation and final pH of *Aspergillus ustus* on different sulphur sources.

Sl. No.	*Nitrogen sources*	*Dry Weight*	*Sporu-lation*	*Final pH*
1	Copper Sulphate	000	Nil	7.0
2	Ferrous Sulphate	098	Fair	7.0
3	Potassium Persulphate	094	Poor	7.0
4	Potassium Metabisulphite	042	Fair	7.5
5	Thiourea	055	Fair	6.0
6	Ammonium Sulphate	068	Fair	7.0
7	Magnesium Sulphate	090	Good	7.0
8	Sodium Sulphate (Anhy.)	066	Fair	7.0
9	Sodium Metabisulphite	074	Good	6.5
10	Sodium Thiosulphate	074	Excellent	7.5
11	No Sulphur	055	Fair	7.0
	G.M.	65.09		

Summary of dry weight results and conclusion at 5% level of P.

Treatments	:	Highly significant
Replicates	:	Non-significant
S.E.	:	4.3251
C.D. at 5% level of P.	:	$\pm$ 8.9703
Dry weight results	:	$\overline{2\ 3\ 7} > \overline{9\ 10\ 6\ 8} > \overline{5\ 11} > 4 > 1$

Table - 30 : Average dry weight (in mg), sporulation and final pH of *Chaetomium cymbiforme* on different sulphur sources.

Sl. No.	*Nitrogen sources*	*Dry Weight*	*Sporu-lation*	*Final pH*
1	Copper Sulphate	028	Nil	4.0
2	Ferrous Sulphate	088	Fair	7.0
3	Potassium Persulphate	087	Fair	6.5
4	Potassium Metabisulphite	032	Poor	7.5
5	Thiourea	050	Fair	6.0
6	Ammonium Sulphate	071	Good	7.0
7	Magnesium Sulphate	110	Good	7.0
8	Sodium Sulphate (Anhy.)	090	Poor	6.5
9	Sodium Metabisulphite	093	Fair	6.5
10	Sodium Thiosulphate	090	Fair	7.5
11	No Sulphur	058	Fair	7.0
	G.M.	72.45		

Summary of dry weight results and conclusion at 5% level of P.

Treatments	:	Highly significant
Replicates	:	Non-significant
S.E.	:	4.5946
C.D. at 5% level of P.	:	$\pm$ 9.5844
Dry weight results	:	$7 > \overline{9\ 8\ 10\ 2\ 3} > 6 > \overline{11\ 5} > \overline{4\ 1}$

In all the cases poor to good sporulation was observed while pH drifted in most of the cases towards neutral side. However, in some cases it drifted towards acidic side or alkaline side.

A perusal of the tables and the above results, clearly revealed that all the four organisms were specific in their growth and sporulation responses towards various sulphur sources and there was no close correlation between the growth and sporulation. It was interesting to note that in certain cases, when growth was moderate sporulation was found excellent.

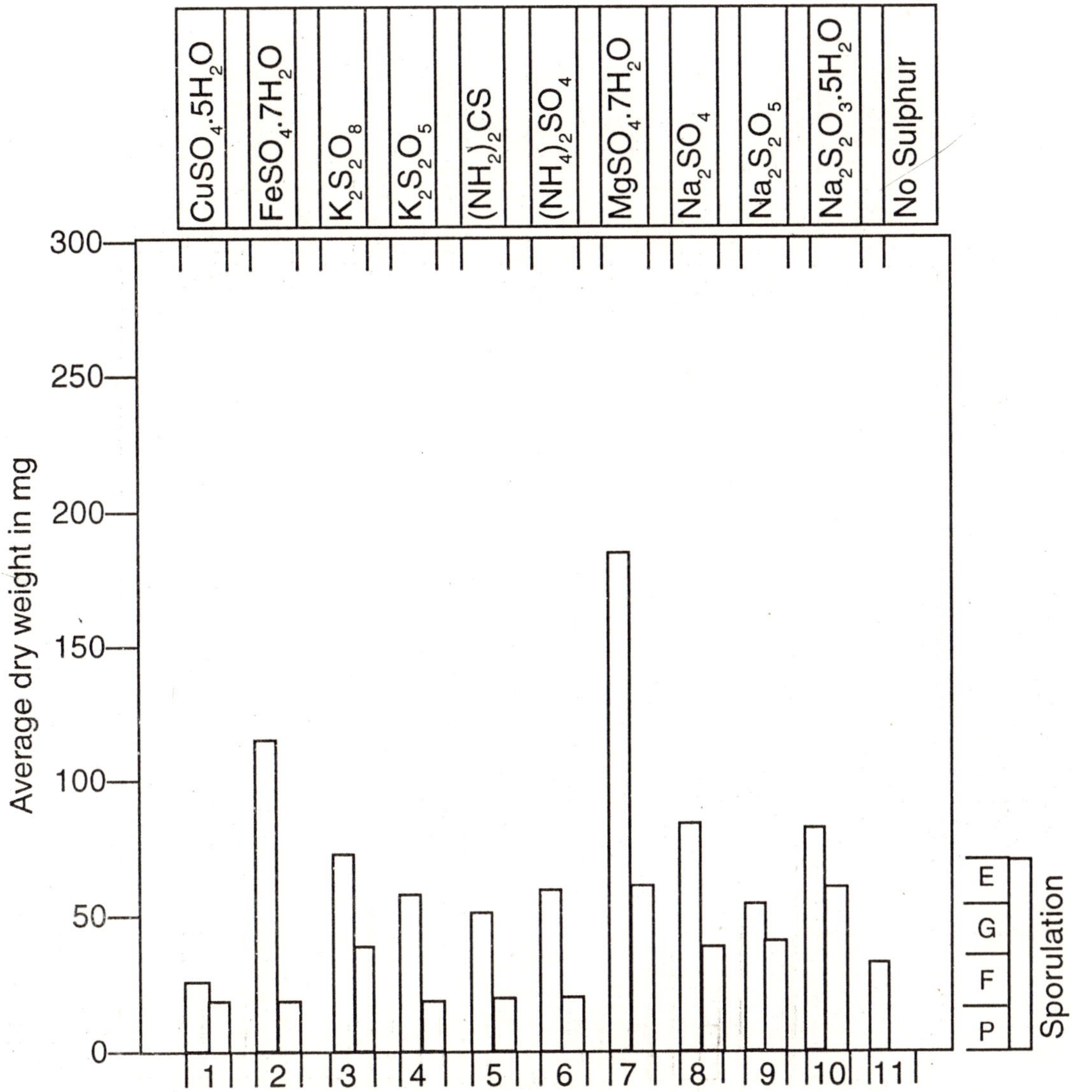

Fig. 23 : Average dry weight and sporulation of *C. verruculosus* on various Sulphur sources.

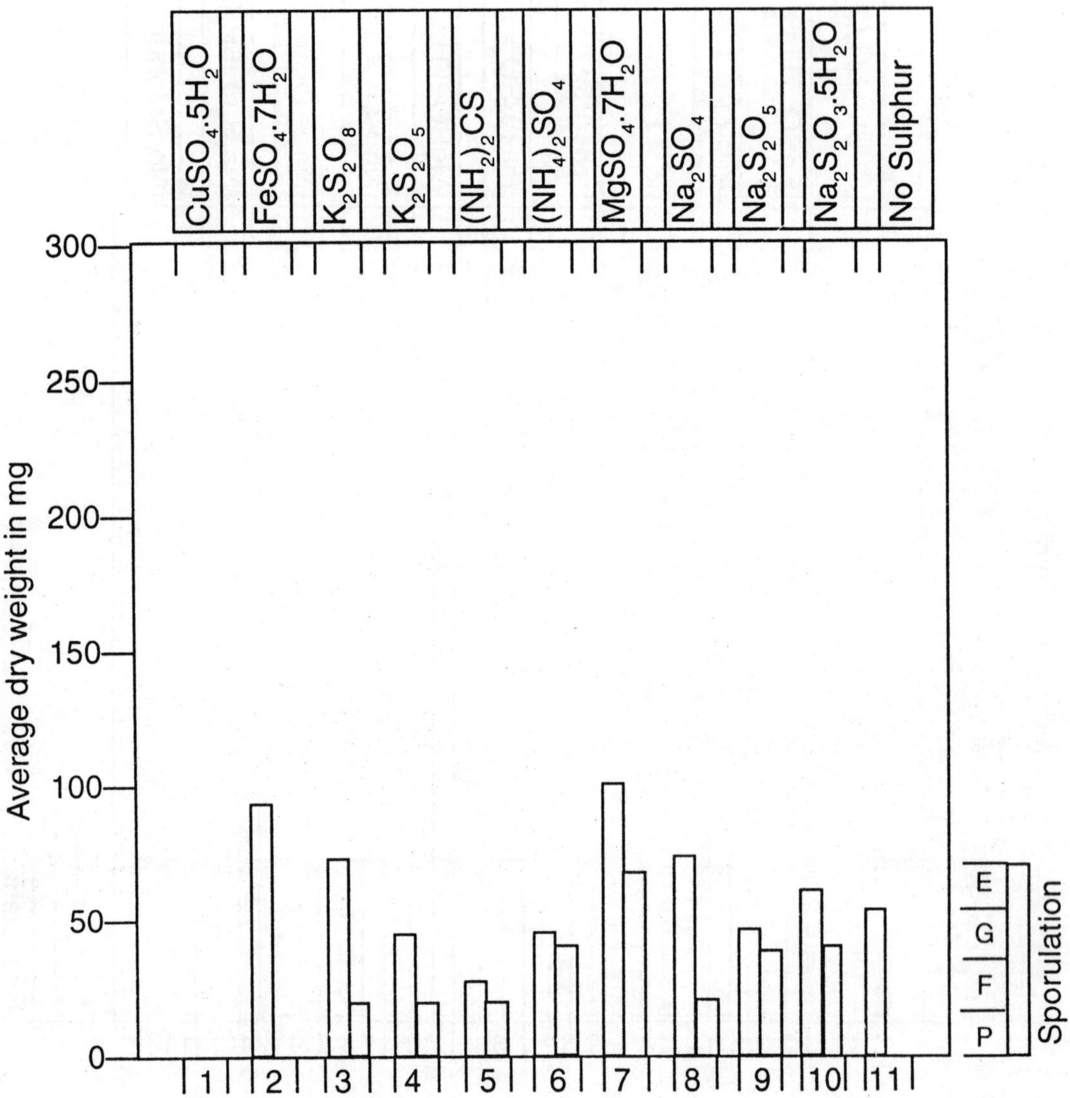

Fig. 24 : Average dry weight and sporulation of *P. citrinum* on various Sulphur sources.

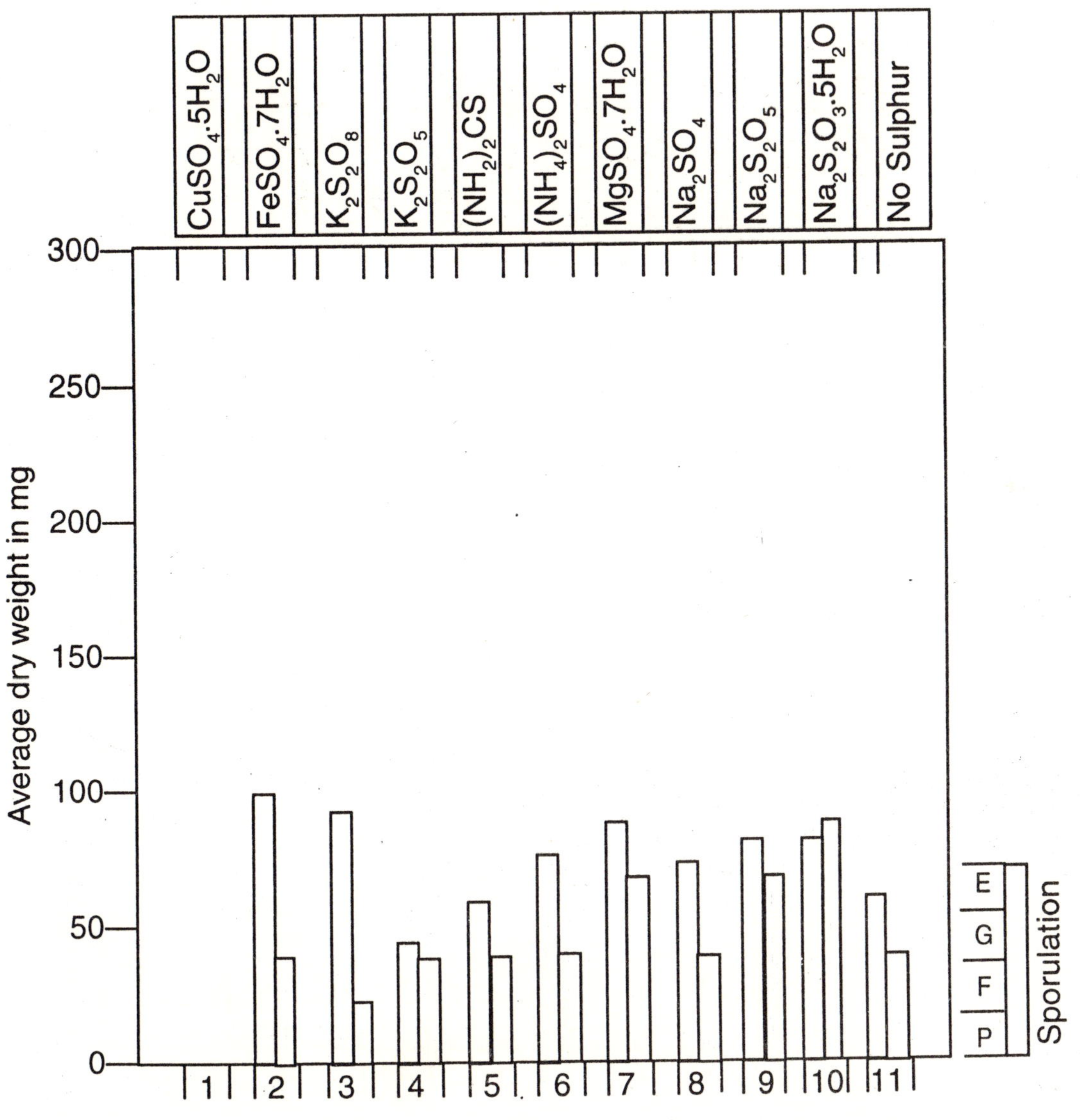

Fig. 25 : Average dry weight and sporulation of *A. ustus* on various Sulphur sources.

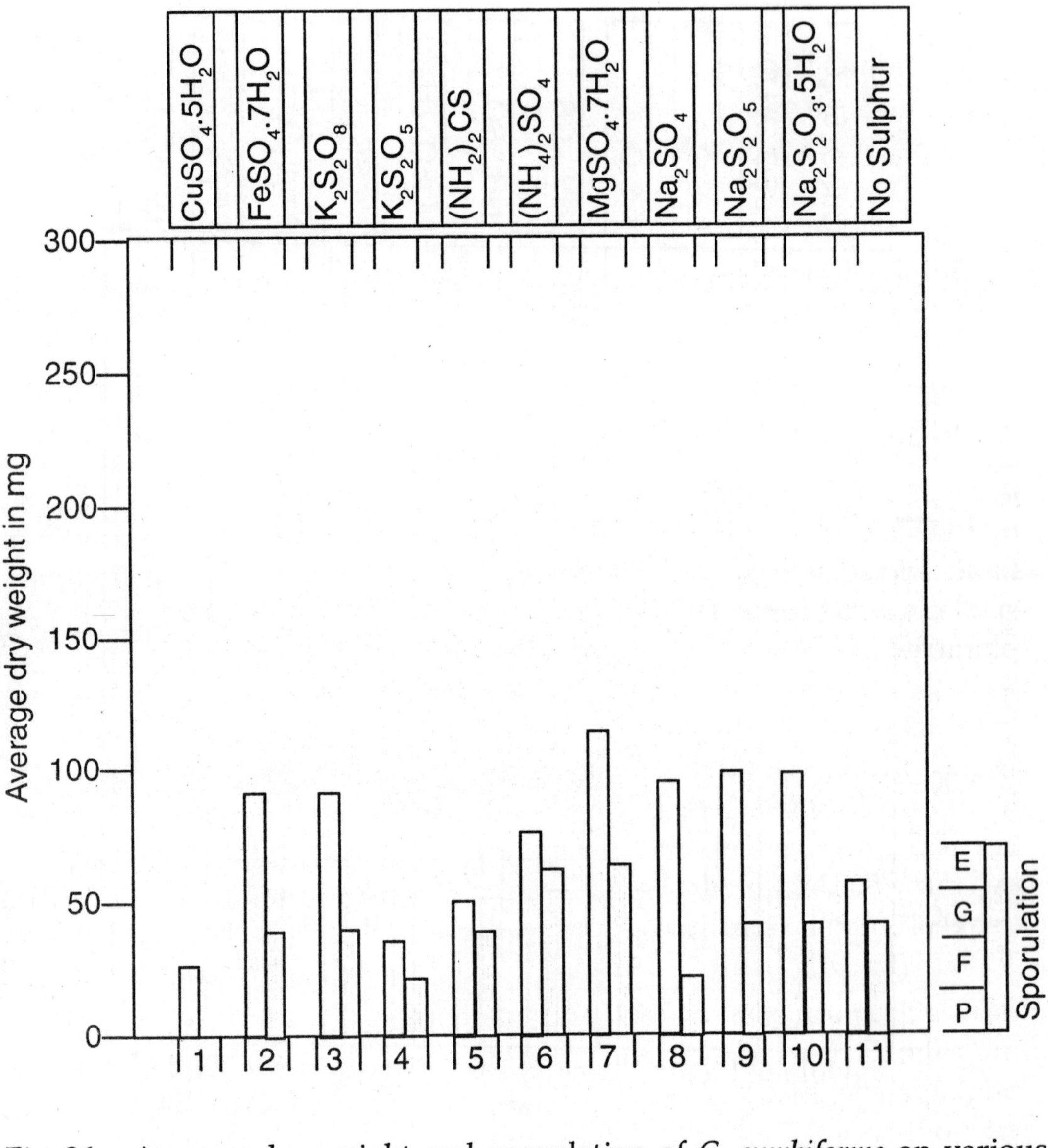

Fig. 26 : Average dry weight and sporulation of *C. cymbiforme* on various Sulphur sources.

5.8 Effect of Different Phosphorus Sources

Phosphorus is essential for all forms of life. Phosphorus compounds play an important role in the functions of chemical transformations and energy transfer (Lilly and Barnett, 1951). Apparently phosphorus is utilized only when it is in the form of phosphate and functions mainly in the form of phosphate esters. Both organic and inorganic phosphates may be used as sources of this element. There exist ample evidences to show the importance of this non-metallic element in the nutrition of fungi. Raulin as early as in 1869 recognised the essentiality of phosphorus in the nutrition of *Aspergillus niger*. Omission of phosphate from synthetic medium reduced the yield approximately to 50%. It is suggested that sugars could not be oxidised and nitrates could not be reduced to amino acids without an adequate supply of phosphorus. If it is not present in sufficient quantity, it may cause several metabolic disturbances (Lilly and Barnett, 1951).

Arya (1990-91 & 1991) stated that phosphours plays an important part in the initial break-down of sugars through the processes of phosphorylation. The amount of phosphorus required is small, as it is not used-up in the process but repeatedly becomes available for further phosphorylation of sugars.

It is reported by Pathak and Dhawan (1989) that the concentration of phosphorus, optimal for certain processes are inhibitory to others.

In the present investigation following sources of phosphorus have been included to determine the growth and sporulation of *C. verruculosus*, *P. citrinum*, *A. ustus* and *C. cymbiforme*.

Potassium hydrogen phosphate dibasic	K_2HPO_4
Potassium dihydrogen phosphate	KH_2PO_4
Sodium phosphate dibasic	$Na_2HPO_4 . 2H_2O$
Sodium dihydrogen phosphate	$NaH_2PO_4 . H_2O$
Ammonium dihydrogen phosphate	$NH_4H_2PO_4$
Sodium hexametaphosphate	$Na(PO_3)_6$
Sodium pyrophosphate	$Na_2P_2O_7 . 10H_2O$

Beside these sources a medium was also prepared without any source of phosphorus. The dry weight results, sporulation and change in pH due to growth of different fungi have been recorded in tables 31 to 34 and shown in figs. 27 to 30.

A perusal of tables 31 to 34 indicated that in absence of phosphorus compound growth was nil in all the four fungi under study. Arya (1990-91 & 1991) reported some growth in four spp. of *Phomopsis*. However, *Fusarium solani, Botryodiplodia ananassae* and *Macrophomina phaseolina* (Bhargava and Tandon, 1963) and three isolates of *Colletotrichum capsici* (Lal, 1977) showed their inability to grow in complete absence of this compound.

Potassium hydrogen phosphate, potassium dihydrogen phosphate and sodium phosphate induced good growth of all the four fungi while sporulation was found good in potassium dihydrogen phosphate. Arya (1990-91 & 1991) also reported the same results in the case of *Phomopsis* spp. Potassium phosphate enhanced the growth of *A. niger* (Pericin *et al.*, 1992).

Sodium dihydrogen phosphate showed good growth of *A. ustus* and *C. cymbifrome* while moderate growth of *C. verruculosus* and *P. citrinum*. Sporulation was found excellent in *P. citrinum*, good in *A. ustus, C. cymbiforme* and nil in *C. verruculosus*. Arya (1982) stated that this compound is a moderate source of phosphorus for *Phomopsis viticola* and *P. psidii*. However, it was found as a good source of phosphorus for *P. sapotae*. (Rai, 1982). Dibasic potassium phosphate was found best as a phosphorus source for *Alternaria brassicae* (Ansari *et al.*, 1989).

Ammonium dihydrogen phosphate, sodium hexametaphosphate and sodium pyrophosphate are the sources of phosphorus which induced good to moderate growth of all the four fungi and exhibited varying degree of sporulation. Arya (1982) reported ammonium dihydrogen phosphate as a poor source of phosphorus for *Phomopsis* spp., while Rai (1982) reported the same source as a moderate source for *P. sapotae*.

Due to growth of different organisms the pH of the medium drifted towards neutrality and in some cases it was found alkaline.

The above results clearly indicate wide variations among the four fungi regarding their choice for different phosphorus compounds. It was noticed that in some cases growth of organisms was found good while sporulation was poor to fair. Therefore, there is no relationship between the growth and sporulation. It is also concluded that presence of phosphorus compounds is essential for good growth and sporulation of fungi.

Table - 31 : Average dry weight (in mg), sporulation and final pH of *C. verruculosus* on various phosphorus sources.

Sl. No.	*Phosphorus sources*	*Dry Weight*	*Sporu-lation*	*Final pH*
1	Potassium hydrogen phosphate dibasic	190	Poor	8.0
2	Potassium dihydrogen phosphate	185	Good	6.0
3	Sodium phosphate dibasic	124	Poor	8.0
4	Sodium dihydrogen phosphate	098	Nil	7.5
5	Ammonium dihydrogen phosphate	112	Nil	2.0
6	Sodium hexametaphosphate	108	Excellent	7.5
7	Sodium pyrophosphate	103	Nil	7.0
8	No phosphorus	000	Nil	5.5
	G.M.	115		

Summary of dry weight results and conclusion at 5% level of P.

Treamtment	:	Highly significant
Replicates	:	Non-significant
S.E.	:	10.6845
C.D. at 5% level of P.	:	$\pm$ 22.65114
Dry weight results	:	$\overline{1\ 2} > \overline{3\ \underline{5\ 6\ 7}}\underline{\ 4} > 8$

Table - 32 : Average dry weight (in mg), sporulation and final pH of *P. citrinum* on various phosphorus sources.

Sl. No.	*Phosphorus sources*	*Dry Weight*	*Sporu-lation*	*Final pH*
1	Potassium hydrogen phosphate dibasic	110	Good	8.0
2	Potassium dihydrogen phosphate	099	Good	6.0
3	Sodium phosphate dibasic	114	Excellent	7.0
4	Sodium dihydrogen phosphate	084	Excellent	7.0
5	Ammonium dihydrogen phosphate	072	Excellent	6.0
6	Sodium hexametaphosphate	080	Excellent	6.5
7	Sodium pyrophosphate	103	Excellent	5.0
8	No phosphorus	000	Nil	3.5
	G.M.	82.7		

Summary of dry weight results and conclusion at 5% level of P.

Treatments	:	Highly significant
Replicates	:	Non-significant
S.E.	:	5.3255
C.D. at 5% level of P.	:	$\pm$11.2900
Dry weight results	:	$\overline{3\,\underline{1\,7\,2}} > \overline{4\,\underline{6\,5}} > 8$

Table - 33 : Average dry weight (in mg), sporulation and final pH of *A. ustus* on various phosphorus sources.

Sl. No.	*Phosphorus sources*	*Dry Weight*	*Sporu-lation*	*Final pH*
1	Potassium hydrogen phosphate dibasic	106	Fair	8.0
2	Potassium dihydrogen phosphate	090	Good	5.5
3	Sodium phosphate dibasic	136	Fair	8.0
4	Sodium dihydrogen phosphate	123	Good	7.0
5	Ammonium dihydrogen phosphate	090	Fair	6.5
6	Sodium hexametaphosphate	088	Excellent	7.0
7	Sodium pyrophosphate	122	Poor	7.0
8	No phosphorus	000	Nil	5.0
	G.M.	94.4		

Summary of dry weight results and conclusion at 5% level of P.

Treatments	:	Highly significant
Replicates	:	Non-significant
S.E.	:	4.988
C.D. at 5% level of P.	:	$\pm$10.5763
Dry weight results	:	$3 > \overline{4\ 7} > 1 > \overline{2\ 5\ 6} > 8$

Table - 34 : Average dry weight (in mg), sporulation and final pH of *C. cymbiforme* on various phosphorus sources.

Sl. No.	*Phosphorus sources*	*Dry Weight*	*Sporu-lation*	*Final pH*
1	Potassium hydrogen phosphate dibasic	130	Good	6.0
2	Potassium dihydrogen phosphate	110	Good	7.5
3	Sodium phosphate dibasic	142	Good	7.0
4	Sodium dihydrogen phosphate	108	Good	7.5
5	Ammonium dihydrogen Phosphate	078	Excellent	6.5
6	Sodium hexametaphosphate	081	Excellent	7.5
7	Sodium pyrophosphate	138	Fair	5.0
8	No phosphorus	000	Nil	3.5
	G.M.	98.4		

Summary of dry weight results and conclusion at 5% level of P.

Treatments	:	Highly significant
Replicates	:	Non-significant
S.E.	:	8.0966
C.D. at 5% level of P.	:	± 16.8895
Dry weight results	:	$\overline{3\;7\;1} > \overline{2\;4} > \overline{6\;5} > 8$

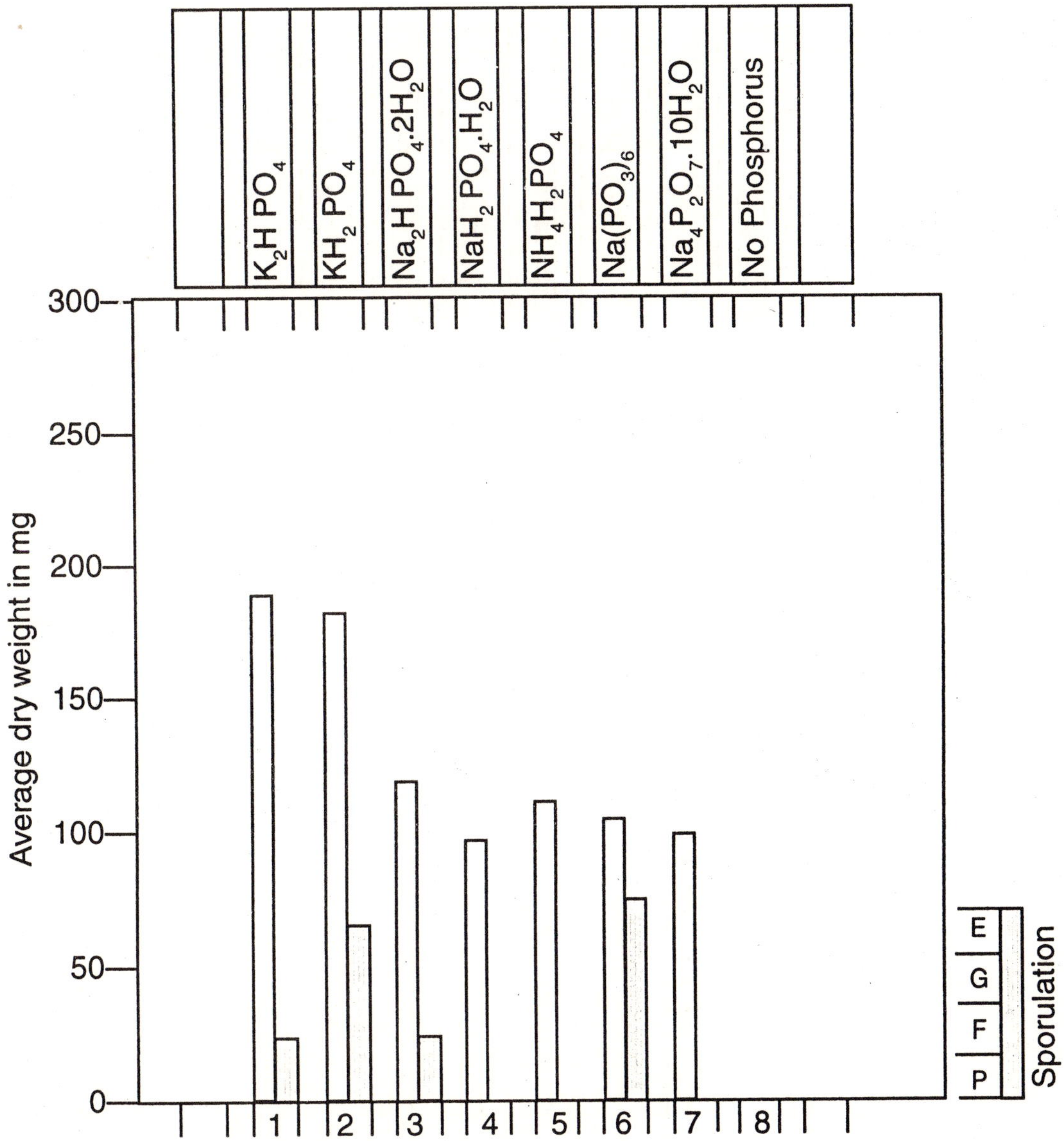

Fig. 27 : Average dry weight and sporulation of *C. verruculosus* on various Phosphorus sources.

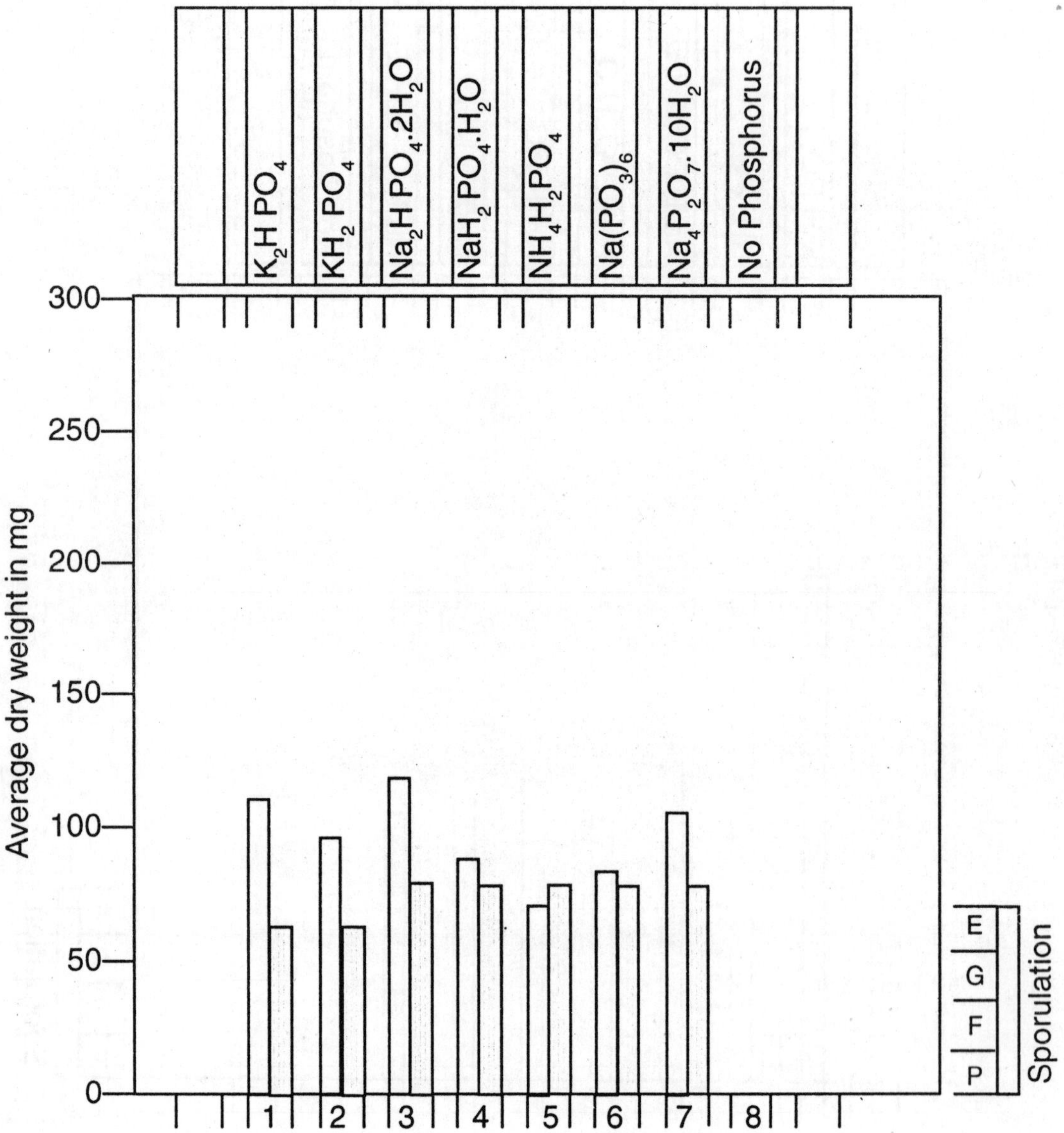

Fig. 28 : Average dry weight and sporulation of *P. citrinum* on various Phosphorus sources.

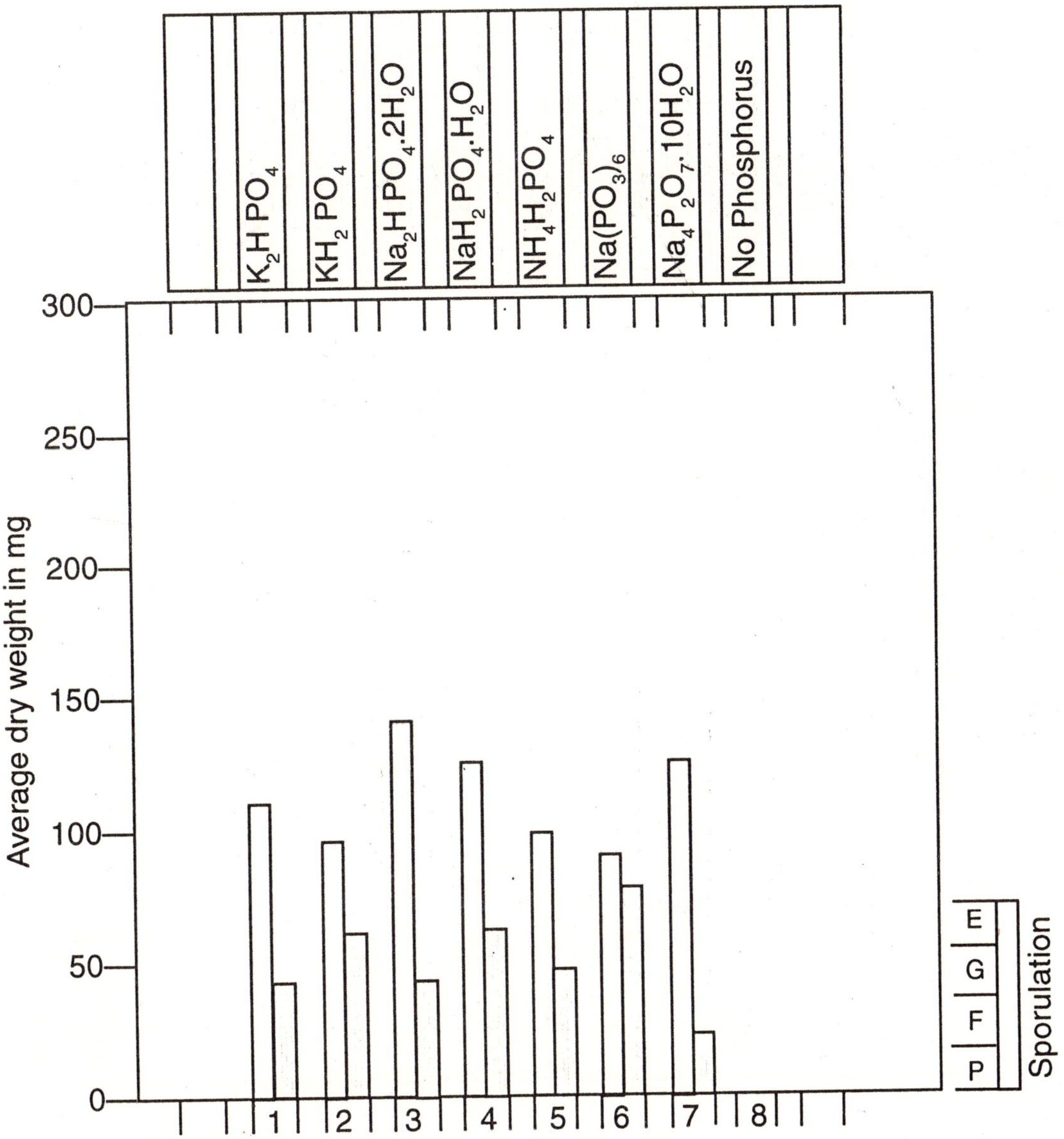

Fig. 29 : Average dry weight and sporulation of *A. ustus* on various Phosphorus sources.

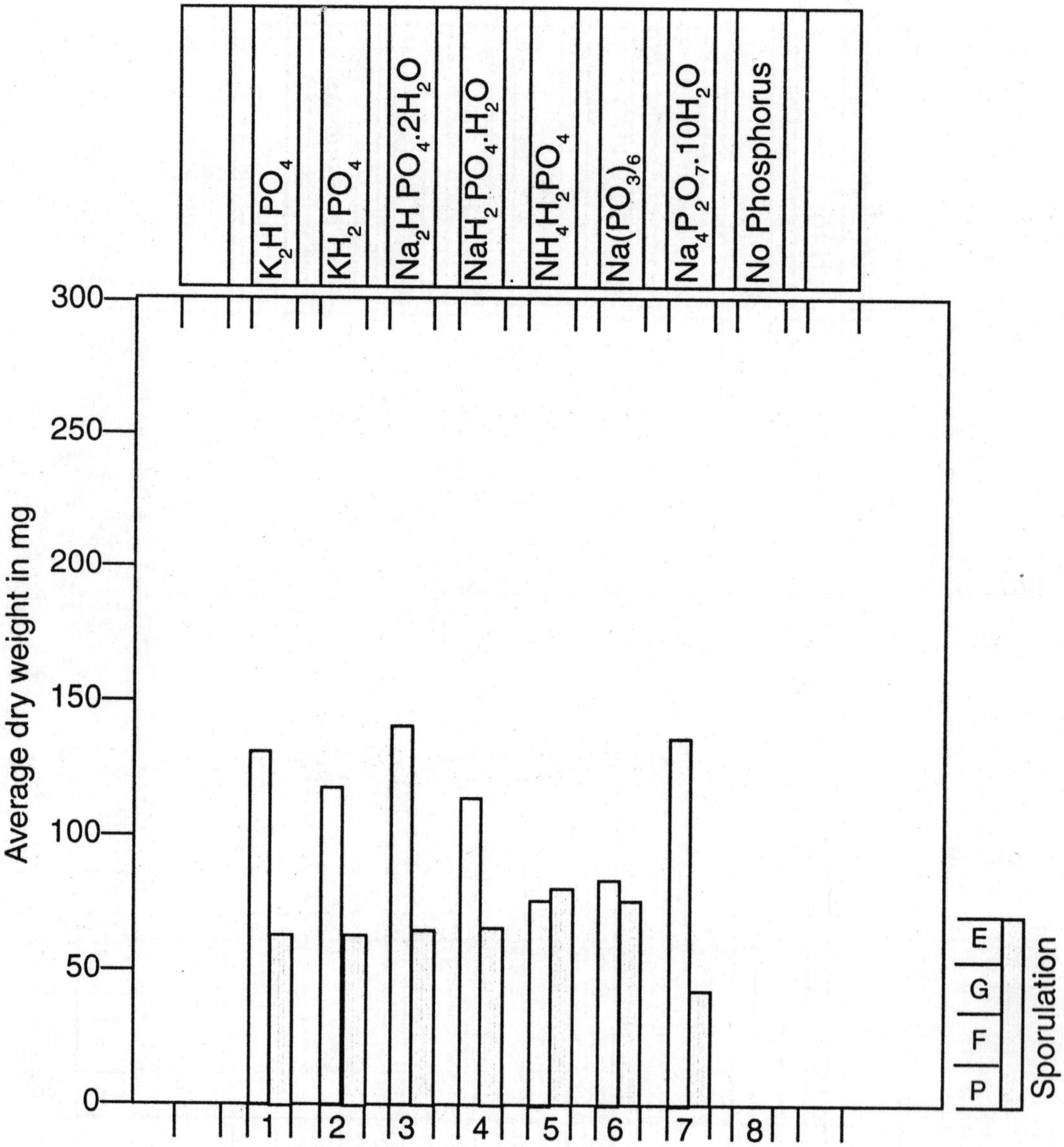

Fig. 30 : Average dry weight and sporulation of *C. cymbiforme* on various Phosphorus sources.

Chapter–6

Aeromycoflora and Biodeterioration Studies

6.1 Studies on Indoor Aeromycoflora

Biological pollution caused by microbes has been a problem since time immemorial. Spores of fungi, bacteria and actinomycetes are always present in the atmosphere. In a tropical country like India, the problem of biodeterioration of museum objects is quite acute. Museum is a public place. Several visitors visit the museum during the day time. In most of the museums visitors are allowed to enter inside the galleries with shoes. This brings microorganisms into the museum premises along with dust. Another source of microorganisms inside the museum building is the indoor air which is usually exchanged fairly and rapidly by ventilation with the outside air. As a result of this, the microbial content of air tends to change; however, indoor pollen and spore number is always very much lower than the outside air. But added to microbes coming in through the doors, windows and ventilators, the indoor air commonly has other microbes derived from indoor sources (Tilak, 1991).

The fungal spores present in the air will not cause damage only to museum objects but many of them are potential allergens to human beings. Thus aerospora forms a normal component of the earth's atmosphere.

Tilak (1991) studied the indoor aeromycoflora of Ajanta caves to study the problem of biodeterioration of cave paintings. A study of aerospora has been done by Singh and Tandon (1991) at Bharat Kala Bhawan to study the problems of biodeterioration in the museum. Srivastava (1991) has studied aeromycology of *Chaetomium* as this fungus has been frequently reported from various habitats. Earlier Greathouse and Ames (1945) have described fabric deterioration by 16 species of *Chaetomium.* This organism being a cellulolytic fungus, causes irreparable damage to paper materials including consolidation of books in libraries. Therefore, they suggested the study of aeromycology of library environment to overcome such types of damages. Similar studies have been done by Strzelczyk and Leznicka (1981) from Poland. According to Kowalik (1980) development of fungi on paper depends on pH of the object.

Chaetomium sp., *Trichoderma viride* and *T. ressei* are cellulolytic fungi and contribute significantly to paper damage (Szczepanowska, 1986). Dhawan and

Agrawal (1986) have studied fungal flora of miniature paper painting and lithographs of the state museum, Lucknow and isolated 23 fungal species. Arai (1987) isolated 7 strains belonging to the absolute tonophilic fungi from foxing. Jeyaraj (1991) found *Ascotricha charterum* and *Aspergillus niger* on books. Laxmikant and Mathur (1991) reported 33 different fungi from manuscripts, books, wood-crafts, paintings and wall hangings.

Studies on aerospora have been done in different cities like Nagpur, Delhi and Kanpur by Chitaley and Bajaj (1973), Agrawal *et al.* (1969) and Nigam & Pathak (1991) respectively. Mycoflora present inside a room is entirely different from outside and is a new concept.

Thus, aerobiological studies have been assuming awareness due to their application in various fields. These studies are helpful in assessing the deterioration and management of environment. The aerobiological data may be used for the formulation of diseases forecasting system of plants (Mishra & Mishra, 1992) and human beings (Davies and Smith, 1973; and Spieksma, 1980).

During the study, it was a common observation that numerous fungi were repeatedly isolated from treated and untreated objects. It immediately raises a question - "How healthy are our museums?" (Fig. 31). And needs some strong actions to prevent the biodeterioration of valuable cultural property.

Indoor aeromycoflora of the following institutions was studied.

1. Museum and Picture Gallery, Baroda.
2. Department of Museology, M.S. University, Baroda.
3. Oriental Institute, M.S. University, Baroda.
4. Smt. Hansa Mehta Library, M.S. University, Baroda.

Both qualitative and quantitative estimations of fungi were done. Culture plate exposure method (Hyde and Williams, 1961) has been used as gravity slide exposure method is found to have following draw-backs (Nigam and Pathak, 1991) :

(i) Similar spores are not identified upto generic level (e.g. *Aspergillus, Penicillium,* and *Mucor* etc.)

(ii) Some spores are too small or too transparent or do not have sufficient individual characteristics e.g. Yeast and *Phoma* etc.

(iii) Fungal hyphae are not identified.

(iv) Many spores cannot be trapped due to strong wind.

INDIAN EXPRESS BARODA 3

City & State

How healthy are our museums?

By N. P. Krishna Kumar

Plea to ensure safe travel for woman passengers

Honarot – an absorbing play

Age takes toll on 'Ajanta of South'

Precious paintings withering at 'art gallery'

By Tanvir Siddiqui

Microbes defacing Ajanta paintings

NEW DELHI • THE HINDUSTAN TIMES • WEDNESDAY NOVEMBER 2 1994

HEALTH

Airborne fungi in Delhi hospitals

Over a hundred fungal forms have been identified in the wards — some are a menace to high-risk patients, warns N. C. Jain

Paintings wither at 'gallery'

Fig. 31 : News paper cuttings reporting role of microbes as biopollutants.

1. Indoor Aeromycoflora of Museum and Picture Gallery, Baroda :

A mycological analysis of air in Museum and Picture Gallery was done at 6 places i.e. Indian Art Gallery, Nepalese and Tibetan Galleries, Islamic, Picture, Ethnological Galleries and Laboratory (Fig. 32).

In these studies colonies of 22 fungal species belonging to 12 genera were recorded. Of these 1 belongs to Zygomycotina, 11 to Ascomycotina and 10 to Deuteromycotina group. Of the six places where petriplates were kept, picture gallery and Indian Art gallery harboured maximum (175 and 162) colonies while in Nepalese Gallery and Laboratory there were least (102 and 114) colonies. Of all these species *Aspergillus niger* and *Penicillium citrinum* constituted about 10% and 8% of the total colonies respectively (Tab. No. 35).

Isolations from exposed petriplates resulted in pure culture of 5 *Aspergillus* and 2 *Penicillium* species. One perfect state of *Aspergillus* i.e. *Emericella nidulans* was also isolated. These are common species causing biodeteriration of various museum objects. The maximum and minimum relative humidity in the month of isolation was 52% and 43% respectively and average temperature was 25.6^{0}C.

2. Indoor aeromycoflora of the Department of Museology, M.S. University, Baroda :

Table 36 shows that spores of 12 different fungi were present in the department. Maximum number of colonies were obtained in the museum of the department located on the ground floor while least number of fungal species were obtained on second floor. Among these fungi, occurrence of *Chaetomium globosum, Monoascus ruber* and *Syncephalastrum racemosum* is feature of special interest.

3. Indoor aeromycoflora of Oriental Institute, Baroda :

Table 37 shows that only 4 fungi were isolated in petriplates exposed in Oriental Institute. 29 colonies of 4 different fungi i.e. *Aspergillus* sp., *A. niger, Cochliobolus verruculosus* and *Emericella veriecolor* were obtained.

4. Indoor aeromycoflora of Smt. Hansa Mehta Library, Baroda :

Similar type of study was undertaken in Smt. Hansa Mehta Library to findout the fungal spores present in the air of the building. 54 colonies of 7 different fungal genera were obtained (Tab. 38).

Plant Pathogenic Fungi :

In the present investigation many fungi were found to be plant pathogens such as *Alternaria alternata, Cladosporium herbarum, Fusarium moniliforme, F. oxysporum* and *Rhizopus* sp.

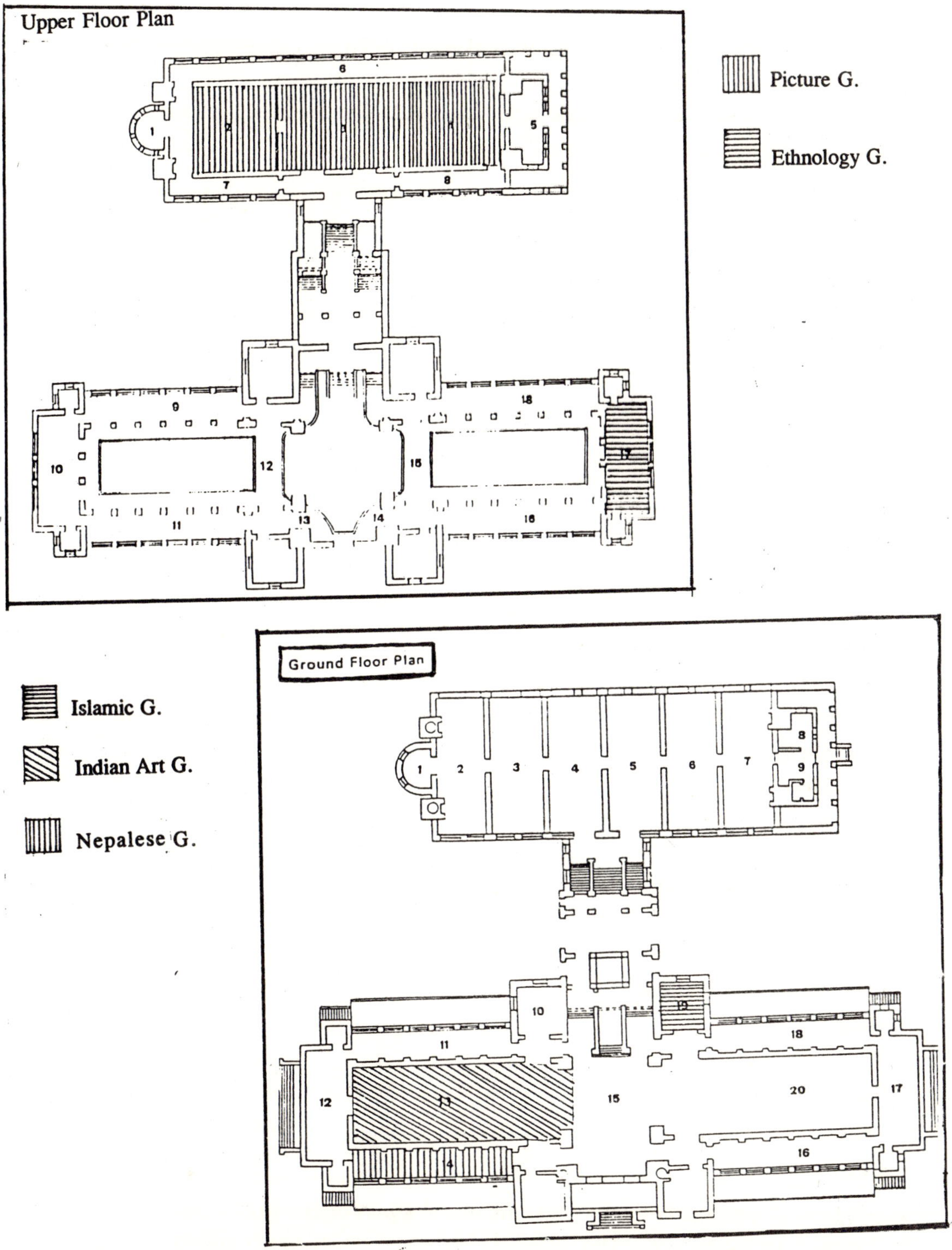

Fig. 32 Lay-out plan of the Museum and Picture Gallery, Baroda.

Table - 35 : Indoor aeromycoflora in different galleries of Museum and Piture Gallery, Baroda.

S.No.	Fungi	IA	NG	IG	PG	EG	Lab.	Total	%
1.	*Alternaria alternata* (Fr.) Keisslar	5	-	-	-	-	5	10	1.22
2.	*Aspergillus flavus* Link.	5	5	1	5	10	10	36	4.42
3.	*A. niger* van Tieghem	10	20	-	20	10	20	80	9.42
4.	*A. parasiticus* Spears	10	10	10	-	-	10	40	4.91
5.	*A. sydowii* (Bain & Sart) Thom. & Church	5	5	10	10	5	5	40	4.91
6.	*A. versicolor* (Vuill.) Tiraboschi	-	-	10	20	10	5	45	5.52
7.	*Cladosporium herbarum* (Pres.) Link.	2	2	5	-	-	-	09	1.10
8.	*Chaetomium gangligerum* L. Ames	-	-	10	10	-	-	20	2.45
9.	*C. globosum* Kunze and Schm.	5	-	-	-	-	10	15	1.84
10.	*C. indicum* Corda	-	-	-	10	-	2	12	1.47
11.	*Curvularia lunata* (Wakker) Boedijn	15	-	10	5	10	2	42	5.15
12.	*Emericella nidulans* (Eldam) Vuilla.	10	-	5	10	-	5	30	3.68
13.	*Fusarium moniliforme* J. Sheld.	10	10	-	-	10	-	30	3.68
14.	*F. oxysporum* Schlecht.	10	-	20	5	10	-	45	5.52
15.	*Fusarium* sp.	5	5	-	-	-	-	10	1.22
16.	*Gliocladium* sp.	5	-	10	-	-	-	15	1.84
17.	*Monilia sitophila* (Mont.) Sacc.	10	5	5	5	-	-	25	3.07
18.	*Penicillium citrinum* Thom.	20	5	20	20	-	-	65	7.98
19.	*P. nigricans* (Bain) Thom.	10	-	10	10	10	10	50	6.14
20.	*Rhizopus stolonifer* (Ehr. ex Fr.) Lind.	10	20	-	20	10	-	60	7.37
21.	*Trichoderma lianorum* (Tode) Harz.	-	-	-	10	5	10	20	2.45
22.	*T. viride* Pers. ex Fr.	-	-	-	-	10	10	20	2.45
23.	Sterile white mycelium	10	10	10	5	5	5	45	5.52
24.	Sterile black mycelium	5	5	10	10	10	5	45	5.52
	Total	162	102	146	175	115	114	814	

Note: The data represent the total colonies produced after 7 days on 5 petriplates.
IA = Indian Art gallery; NG = Nepalese gallery; IG = Islamic gallery;
PG = Picture gallery; EG = Ethnological gallery; Lab. = Conservation laboratory.

Table-36 : Indoor aeromycoflora at different places in the Deptt. of Museology, Baroda .

Sl. No.	Fungi	Total colonies produced in 5 petriplates				
		Museum	Firest Floor	Second Floor	Total	%
1.	*Chaetomium globosum* Kunze and Sohm.	5	-	2	07	5.5
2.	*Aspergillus fumigatus* Fresen.	5	5	-	10	7.8
3.	*A. niger* van Tieghem	10	15	10	35	27.5
4.	*A. ustus* (Bainier) Thom. & Church	5	2	1	08	6.2
5.	*Cochliobolus pallescens* (Tsuda & Ueyama) Sivan	5	5	-	10	7.8
6.	*C. verruculosus* (Tsuda & Ueyama) Sivan	-	2	-	02	1.5
7.	*Emericella variecolor* Berk. & Broome	5	5	5	15	11.8
8.	*Monoascus ruber* van Tieghem	2	-	-	02	1.5
9.	*Penicillium* sp.	5	5	2	12	9.4
10.	*P. griseofulvum* Dierckx	2	5	5	12	9.4
11.	*Rhizopus* sp.	5	2	5	12	9.4
12.	*Syncephalastrum racemosum* Cohn ex Schroter.	2	-	-	02	1.5
	Total 12 spp.	51	46	30	127	

Table - 37 : Indoor aeromycoflora of Oriental Institute, Baroda .

Sl. No.	Fungi	Total colonies produced in 5 Petriplates	%
1.	*Aspergillus* sp.	5	17.2
2.	*A. niger* van Tieghem	12	41.3
3.	*Cochliobolus verruculosus* (Tsuda & Ueyama) Sivan	2	6.8
4.	*Emericella veriecolor* Berk.. & Broome.	10	34.4
	Total 4 Spp.	29	99.7

Table - 38 : Indoor aeromycoflora of Smt. H. M. Library, Baroda .

Sl. No.	Fungi	Total colonies produced in 5 Petriplates	%
1.	*Aspergillus flavus* Link.	5	9.2
2.	*A. niger* van Tieghem	15	27.5
3.	*A. fumigatus* Fresen.	7	13.0
4.	*Chaetomium gangligerum* L. Ames	3	5.5
5.	*Chaetomium globosum* Kunze and Sohm.	4	7.4
6.	*Penicillium* sp.	10	18.5
7.	*Rhizopus* sp.	10	18.5
	Total 7 Spp.	54	

Their occurrence may be due to surrounding environment full of plants. The fungal spores may be blown from these sources and spread in the museum environment. It is interesting to note that some fungi like *Cochliobolus pallescens, Phoma* sp., *Fusarium semitectum* (Shah *et al.,* 1991); *Cladosporium herbarum, Aspergillus niger* (Shah *et al.,* 1992); *Pencillium citrinum* (Shah *et* al., 1992-93) and *Cochliobolus lunatus* (Shah & Arya, 1995) were found to cause deterioration of museum objects in Baroda.

Fungi is a major contributor to the deterioration of cultural property in the tropical countries. As recorded in the results the spore diversity was very high. There was a dominance of the members of Deuteromycotina in all the observations.

The presence of these fungal inocula in the atmosphere clearly warrants immediate protective measures and reduction of humidity levels in the galleries to avoid fresh attacks on museum materials.

6.2 Biodeterioration of Different Woods

Wood is one of the most important natural resources in the world today and it is likely to have been of importance to man from the earliest times. It is difficult to trace the history of wood for making art objects but it is certain that this must have been one of the first materials being used for the purpose. Historians tell, people may have used wood even before they used stone. Stone being an inorganic material, tools were preserved while wood being a cellulosic/lignolic, chances of its survival are almost nil. "The oldest wooden implement yet known is a spear tip found preserved in peaty deposits at Clacton in Eastern England. This is probably 300,000 years old. According to Dr. Peter Andrews of the Natural History Museum, London this wood is obtained from yew tree. Another ancient spear, from Lehringen in Germany, also seems to have been a thrusting type, though it is less robust than the Clacton one. The spear, which belongs to the last interglacial period, about 1,20,000 years ago, was actually found piercing through the ribs of an elephant skeleton. This spear was badly damaged and it was not worth retrieving" (Gowlett 1984).

"Vital evidence for much wider use of wood in Acheulean comes from Prof. Clark's excavations at Kalambo falls on borders between Zambia and Tanzania. This site was once estimated to be as late as 60,000 years old, but recemization dates suggest it to be about 2,00,000 years old. It is an exceptional site because a large quantity of organic material was preserved permanently waterlogged" (Gowlett 1984).

The earliest archaeological evidence comes of a wooden coffin box, excavated from Harappa, is datable to 2,000 B.C., an Indus Valley Civilization Site. In India, the earliest records of surviving wood has been reported from Chaitya cave at Bhaja in Western India, where two ribs have Brahmi inscriptions engraved in wood. The date of inscription on paleographic grounds appears to be not later than 2nd cen. B.C. (Desphande 1959). The use of wood for temple architecture is described in detail in Vishnudharmottara, a text of the 6th cen. A.D. (Shah 1961).

Thus from prehistoric times wood has been used as convenient multipurpose material for household use like fuel, furniture and antiquities. The wood is extensively used in museum collections particularly in the ethnographical and folk exhibits.

Chemically wood is lignin but small amount of cellulose may be present. In presence of high relative humidity bacteria, fungi and actinomycetes are capable of attacking the wood. In tropical humid climate, fungi constitute a major threat to the preservation of lignin materials. More than 100 varieties of fungi attack wooden materials.

Spores and growing mycelia are mostly responsible for altering asthetic appearance of an art object by creating residues. On surface of the object, mycelia can form dry and brittle crusts, wrinkled and corrugated colonies and sometimes loose, elastic, fluffy structures. It burrows into the substratum and digest it, causing structural damage. The process of degradation involves depolymerisation of lignin and cellulose by enzymes produced by the mycelium. Colours produced by the vegetative mycelium cause staining of the object.

Wood rotting and wood staining fungi are the major agents of microbial deterioration in wooden cultural properties.

Wood Rotting Fungi :

They decompose cellulose and cause darkening of the wood. The fungi in addition, breakdown the lignin and cause the wood to become pale and reduce it to a fibrous mass.

Wood Staining Fungi :

They are characterised by pigmentation of the hyphal wall which causes discolouration of the infected sap wood. This is sometimes blue black in colour. They are able to penetrate through cell wall in a characteristic manner affecting the interior of the wood and growing by decomposing and absorbing the wooden elements decreasing the strength of the wood.

Many scientists have reported a number of fungi causing damage to valuable wooden objects, such as *Aspergillus candidus, A. versicolor, A. flavus, A. fumigatus, A. ochraceus, A. niger, A. nidulans, A. ustus, Penicillium citro-viride, P. dicumbens, P. spiculisporum, P. stecki, Oospora* sp., *Trichoderma viride, Eurotium repens, Fusarium solani, Botryodiplodia* sp., *Syncephalastrum racemosum, Humicola* sp. (Garg and Dhawan 1985-87).

Fungi growing on art objects affect surface characteristic of art work by secreting a variety of fungal pigments and acids. These substances are formed during the metabolic processes. Secreted pigments may stain objects either by direct adhesion or by chemically reacting with the object to yield coloured byproducts. In the latter case result is permanent destruction of the art object.

Fungal growth on wood, deteriorates them by producing different enzymes too. Enzymes are water soluble protein molecules of high molecular weight. They are active in low concentrations and catalyze a particular chemical reaction without being permanently altering in the process. Enzyme production is greatest in brown rots and least in soft rots in which cellulase production is very low (Lyr 1962).

Wood is an ideal calorie carbon source which suffers from the disadvantage that it is highly polymeric and thus highly insoluble. In order to be ingested as a food source by microorganisms, it has to be broken down into small water soluble molecules. It takes long time for breakdown unless a suitable catalyst is present.

Fungal enzyme must be capable to work outside the fungal cell otherwise it cannot make contact with the wood. All wood deteriorating microorganisms use extracellular enzymes which are too large to penetrate into the native cell wall. The moisture in the cell lumen swells, the cellulose thus rendering the internal surface more accessible for the extracellular enzymes to diffuse into the cell wall. These enzymes catalyze the depolymerization of the different cell wall constituents into water soluble and smaller molecular fragments which diffuse back to the fungal lumen where they are assimilated by the fungi (Agrawal *et al.* 1990).

Many timbers such as deodar, rose wood, sisam, sandal wood, sal, babul, teak wood have been widely used in India by the craftsman. Decay of such artifects due to microbial activities has been observed in many museums which results in destruction of the wooden artifects. Therefore, the study has been undertaken to evaluate the effect of four fungi on degradation of certain woods. For this experiment following type of wood were selected.

1. Sal; *Shorea robusta* Gaertn. (Fam. Dipterocarpaceae).
2. Teak; *Tectona grandis* L. (Fam. Verbenaceae).
3. Mango; *Mangifera indica* L. (Fam. Anacardiaceae).
4. Deodar; *Cedrus deodara* (Roxb. ex Lamb.) G. Don; (Fam. Pinaceae).
5. Neem; *Azadirachta indica* A. Juss. (Fam. Meliaceae).

The above woods are not only widely used in making artifects but they are also used to prepare museum furniture for display and storage purposes. Biodeterioration of these materials will certainly destroy the valuable wooden antiquities.

An experiment has been set to observe biodegradation of different woods by four fungi. For this experiment 0.5 g of wooden scrap was mixed with 25 ml of modified Asthana and Hawker's medium 'A'. Details are given in the chapter Investigation Techniques. Percentage loss was calculated in each case and the results are shown in Table - 39.

It is evident from the Table - 39 that the rate of degradation differs in all the cases. Maximum biodegradation was observed in deodar wood in all the fungi while lowest biodegradation was observed in neem wood. *C. cymbiforme* caused equal degradation of neem and mango wood.

Sporulation was good in media containing mango and deodar wood while it varied from poor to excellent in media containing other woods. pH drift was observed towards neutral side.

Lakshmikant and Mathur (1991) reported a number of fungi from Gorakhpur causing damage to wooden materials. This list includes *Penicillium citrinum,* one of the four fungi included in the present study.

Seifert (1983) reported decay of wood by the members of Decrymycetales (Basidiomycotina). He found greater percentage of moisture content in blocks of wood showing more weight loss. The percentage weight loss was directly correlated to percentage change in lignin and cellulose.

6.3 Biodeterioration of Different Papers

Man being social and intellectual personality tried to put his knowledge and experiences for the benefit of future generation. From the earliest times various materials were used for writing purposes such as : clay, stones,

Table 39 : Sporulation, Final pH and Percentage Loss in Dry Weight of Different Woods by Four Fungi.

Sl. No.	Source of Wood	Organisms											
		C. verruculosus			*P. citrinum*			*A. ustus*			*C. cymbiforme*		
		Sporulation	*Final pH*	*% loss*	*Sporulation*	*Final pH*	*% loss*	*Sporulation*	*Final pH*	*% loss*	*Sporulation*	*Final pH*	*% loss*
1	Sal	F	7.0	6.0	G	5.0	14.0	P	6.5	8.4	E	6.0	8.5
2	Teak	P	7.0	6.6	P	5.0	12.0	F	6.5	9.0	G	6.5	13.0
3	Mango	G	6.5	7.0	G	6.5	13.0	G	7.0	8.0	G	6.5	7.4
4	Deodar	G	6.5	12.0	G	6.0	17.6	G	7.0	17.0	G	7.0	15.4
5	Neem	F	7.0	5.0	G	7.0	11.0	G	7.0	7.0	G	7.0	7.4

Note : P = Poor; F = Fair; G = Good; E = Excellent Sporulation.

metals, bones, textiles, papyrus, wood, parchment, vellum, palm leaves, birch bark etc., until paper replaced them. With the advancement of civilization and acquisition of the knowledge of preparing various metals the lead, copper or bronze plates were used for writing.

Sindall (1919) stated that the discovery of wonderful properties of the Egyptian papyrus - *Cyperus papyrus* Linn. (Fam. Cyperaceae) was a great step in development of paper making. Pliny made mention of the writings of Numa, who lived about 670 B.C., whereas Joshi (1989) stated that documents written on *Papyrus* have been found out in Egypt and are as old as 3000 to 2000 B.C.

Earlier Joshi (1947) had reported that parchment or tender skin of animals was also used by Persians and the Greeks for writing purposes. From the beginning of Christian era, palm leaves and bark of trees and other materials were used for writing in various countries in the East as well as in Egypt. Before paper was introduced into India, Bhoj-patra (*Betula utilis* L. bark of the tree, Fam. Betulaceae) was used for writing manuscripts.

The earliest reference of the manufacture of paper is in Chinese encyclopaedia, wherein, it is stated that paper was manufactured in 105 A.D. in China by Tsai-Lun (Sindall 1919, Joshi 1947, Hunter 1957). Later on Arabs learnt the art of paper making from the Chinese prisoners. It is not definitely known when paper making started in India. However, it is believed that Mahmud of Ghaznavi introduced it in the 10th cen. A.D. during his first invasion of India (Hunter 1957). Thus men started to write their evidences on paper itself rather than on any other material.

Paper is one such anthropogenic material which is one of the most essential requirements and is used in different ways. Every year hundreds of millions of tons (about 180 million tons) is produced. Like other cellulosic material it is gradually distintegrated under environmental conditions (Nyuksha 1994).

An ordinary paper used for archival purposes is made up of mainly three materials - cellulose fibres which are chief constituent, rosin or gelatin which is a sizing material and an inorganic inert material like kaolin known as filler.

Chemically cellulose is a polyssacharide chain consisting of a number of glucose molecules. A chain of about 300 to 1000 glucose molecules is normally quite stable and stands up to much physical strains and chemical reactions. The structural formula of cellulose is—

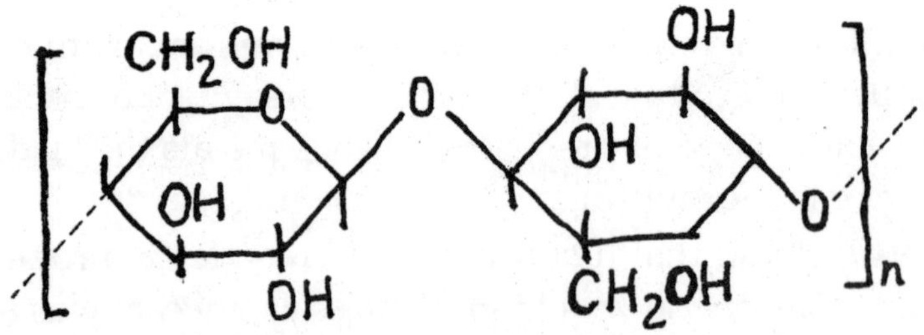

"The deterioration of cellulose, whether by oxidation or hydrolytic processes, is a topic of great importance to the question of the preservation of books and manuscripts. It is relatively stable to autoxidation at ordinary temperature but the primary hydroxyl groups can be progressively oxidised to carboxylic acid groups especially under the influence of light " (Mills and Whilte 1987).

Cellulose is hydrolysed by aqueous acids to glucose units or by hydrolytic enzymes, called cellulases, which are produced by many species of moulds or fungi. These can consequently break down cellulose and utilize the product for their own nutrition (Mills and Whilte 1987).

There are many important documents written in ancient times preserved in Indian Museums which need scientific preservation. Paper being a most sensitive material is prone to many agencies of deterioration and, therefore, scientific preservation of paper documents is a prime need to stop ruins of such an important collection. It is known that in many libraries books are damaged severely by insects, fungi or other ways. In museums these agencies may cause severe and irreparable damage to archival collections. Fungi which induce biodeterioration on paper can be divided into two groups. Those which belong to the first group initiate the degradation process and are mostly saprophytes including representatives of genera *Penicillium, Aspergillus, Alternaria* and *Fusarium*. The second group consists of fungi with adaptivity to definite substrates such as cellulose products *e.g.Trichoderma viride* and *T. reesei*. Most of the members of genus *Chaetomium* are cellulolytic and contribute significantly to paper damage (Szczepańowska, 1986).

Many scientists have worked on biodeterioration of paper materials in different parts of the world. Nyuksha (1956) reported that many species are known which do not take active part in destruction of paper but are found in large quantities such as Mucorales, *Monilia sitophila* and many representatives of *Penicillium* variety (Asymetrica). Strzelczyk and Leznicka (1981) reported consolidation of some books of the division of old prints of the N. Copernicus University Library, Torun. Janposri (1991) reported fungal growth on many ancient paper paintings of tempera technique in Thailand which resulted in flaking of paint and powdering of some of the part of the paintings. Similar observations were also made by several Indian scientists. Dhawan and Agrawal (1986) reported damage to miniature paper paintings and lithographs of the State Museum, Lucknow. Jeyaraj (1991) also observed growth of *Ascotricha charterum, Aspergillus niger* and *Aspergillus* sp. on paper materials. Shah *et al.* (1991) reported *Aspergillus niger, A. sydowii, Syncephalastrum racemosum* and *Penicillium funiculosum* involved in biodeterioration of paper. They further observed that certain chemical treatments (use of Boric acid and Pentachlorophenol etc.), heat treatment and use of ionizing radiations can protect the paper material from microbial attack. Thakre and Bhajbhuje (1991) found many species of fungi destroying books and journals but among these *Aspergillus, Chaetomium thermophile* and *Nigrospora* sp. dominated.

Considering the need of controlling the biodeterioration of paper, it was thought desirable first to study biodegradation of paper caused by four organisms. Papers selected for the experiment were

1. Whatman's Filter Paper No. 1.
2. Common Paper.
3. Brown Paper.
4. News Paper.
5. Old Manuscript.
6. Card Board.
7. Japanese Tissue Paper.
8. Nepalese Tissue Paper.

In a conical flask, 0.5g of paper was taken. 25 ml. of modified Asthana and Hawker's medium 'A' was taken in the same flask as a basal medium. Flasks were autoclaved and then inoculum was incorporated. After incubation for 25 days the fungal mat was harvested by filtering over previously dried Whatman's filter paper no. 1. Actual loss in paper material was determined and then % loss of paper was calculated. Results are shown in Table - 40.

It is evident from the Table - 40 that maximum % loss was observed in Japanese tissue paper by *C. verruculosus* and *A. ustus* whereas, it was maximum in card board by *P. citrinum* and in brown paper by *C. cymbiforme* a member of Chaetomiales group in Ascomycotina. The lowest % loss was observed in common paper by *C. verruculosus* and *A. ustus* while it was minimum in Whatman's paper by *P. citrinum* and in Nepalese tissue paper by *C. cymbiforme*. Japanese and Nepalese tissue papers find extensive application in conservation of museum objects. They are used in restoration of manuscripts and miniature paintings. The results are indicative of less degradation of Nepalese tissue paper as compared to Japanese tissue paper. Out of four fungi three have caused negligible loss of Nepalese tissue paper and hence it can be safely used for restoration work.

Jayanti *et al.* (1992) studied cellulolytic activity of certain fungi. They reported maximum celluloytic activity in terms of filter paper utilization by *Aspergillus flavus* followed appreciably by *Cladosporium cladosporioides, Aspergillus luchyensis* and *Chaetomium globosum*. Verma and Johri (1992) isolated various fungal species to find out the cellulolytic ability with a view to understand their role in cellulose decomposition. The weight loss determination showed considerable variation in cellulolysis of filter paper by different fungal forms. It is presumed that members of Phycomycetes which include Mastigomycotina and Zygomycotina could not degrade filter paper as cellulosic substrate but members of Ascomycotina and Deuteromycotina caused degradation. Lakshmikant *et* al. (1990) reported that *Aspergillus niger, Chaetomium globosum* and *Trichoderma koningii* showed maximum degradation of cellulose at pH 5.0, while *Scopulariopsis brevicaulis* and *Trichoderma roseum* were most active at pH 6.0.

Percentage loss of paper clearly indicates the amount of paper degraded by the fungi. It is observed that rate of biodegradation is different in various types of papers. Biodegradation of paper depends upon physical characteristics such as moisture content, temperature and pH of the paper besides their chemical constituents. Breakdown of the components of the paper takes place due to the enzymes secreted by fungi.

It was also observed that two organisms had a choice for Japanese tissue paper while third one had a choice of brown paper and the another had a choice for card board. Hence, it can be concluded that choice of paper as a food may differ from fungus to fungus.

In all the four organisms sporulation was observed from good to excellent except in *C. verruculosus,* where, it was found fair eventhough percentage loss was observed highest. pH of the medium drifted towards neutral or alkaline side.

Table 40 : Sporulation, Final pH and Percentage Loss in Dry Weight of Different Papers by Four Fungi.

Sl. No.	Types of paper	*Organisms*											
		C. verruculosus			*P. citrinum*			*A. ustus*			*C. cymbiforme*		
		Sporulation	*Final pH*	*% loss*	*Sporulation*	*Final pH*	*% loss*	*Sporulation*	*Final pH*	*% loss*	*Sporulation*	*Final pH*	*% loss*
1	Whatman's filter paper no.1	P	8.0	15.0	E	5.0	11.4	G	7.0	9.5	E	7.0	12.0
2	Common Paper	F	7.0	14.2	E	5.0	48.0	P	7.0	2.5	E	7.0	14.3
3	Brown Paper	F	7.5	37.0	G	6.5	24.0	E	7.5	20.6	G	6.5	35.6
4	News Paper	F	8.0	29.0	E	7.5	16.0	E	8.0	18.8	E	7.0	24.4
5	Old Manuscript	F	7.0	36.0	E	5.0	14.4	F	7.0	15.8	E	7.0	28.4
6	Card Board	F	8.5	44.4	E	8.0	51.4	E	8.5	25.6	G	7.0	22.0
7	Japanese Tissue Paper	F	8.5	49.0	P	8.0	28.0	G	8.0	49.0	G	8.0	22.0
8	Nepalese Tissue Paper	E	7.5	48.0	P	6.5	12.0	G	7.5	20.0	G	7.5	10.0

Note : P = Poor; F = Fair; G = Good; E = Excellent Sporulation.

Chapter—7

Control Studies

Biodeterioration and its control is one of the very important aspects of conservation of cultural property. To avoid the irreparable damage by microbes use of preventive measures and control stratagies are suggested by numerous workers from time to time. Anchabagze (1949) published a report of an investigation of manuscripts at the Georgian State Museum. He enlisted 17 species of fungi and considered air to be the principal cause of the abundant contamination of the material. Omelianskii (1953) reported that after the flood of the river Neva in Colombia in 1924, more than 62,000 books were destroyed due to fungi. Studies on the chemical effects of soft-rot fungi have been limited to *Chaetomium globosum* on beech wood. Soft rot fungi differ from decay fungi in the way they change the wood chemically, similar to white rot fungi in causing a comparatively very little increase in alkali solubility yet behaving like brown rot species in being unable to utilize the wood lignin extensively (Savory and Pinion, 1958; Levi and Preston, 1965).

"In Toscany, in 1966, over 30 libraries and archives suffered extensive damage by fungus after disastrous flood. In Lisbon, in 1967 many valuable manuscripts belonging to the Calouste Gulbenkien were also damaged" (Reed, 1972). In India Narayanan (1963 a and b) reported microbial attack on paper after the flood in Poona in 1961. Dhawan and Agrawal (1986) reported 23 species of fungi from the collection of 1750 miniatures and about 100 lithographs of the State Museum Lucknow. Agrawal (1991) observed some blackish and greyish fungal growth in the paintings received by Lalit Kala Akadmi for exhibition purpose. All this happened within a short span of 4 to 5 days.

Above examples clearly indicate that biodeterioration poses a severe problem to preserve the valuable museum objects. Besides microbes other agents are equally important. Lack of proper attention can cause irreparable damage in different types of organic materials. "A large number of biological entities like algae, insects, mice, rats, pigeons etc. and men as well are responsible for decay of world's valuable heritage (Feilden, 1996)."

Biodeterioration and its control is one of the most important aspects of conservation of cultural property (Agrawal, 1994). The fungal growth can cause,

a) Stains, foxing and discolouration on paper materials.

b) Disfigurement and powdering of the materials.

c) Reduction of the strength of the wood or leather thus reducting their life.

Fungal growth can be seen on a susceptible material if precautionary measures are inadequate. Controlling the fungus on museum objects is concern of an active conservationist. As soon as the presence of fungus is established, fumigation or chemical treatment should be tried. Following methods have been employed to check the fungal growth.

7.1. Preventive measures.

7.2. Remedial measures.

a. Use of Fungicides.

b. Use of Chemicals.

c. Use of Natural Plant Products.

7.1 Preventive Measures :

Aim of this step is to keep the fungal spores away from the objects by some artificial barriers. The measures like good house keeping and proper sanitation, removal of shoes outside the storehouse or gate of a museum, provision of high-speed air curtain and provision of window and door curtains will surely help to avoid fungal attack on museum objects.

Control of environmental conditions like temperature and relative humidity play a very important role in the growth of a fungus. It was observed by Rebrikova and Manturovskay (1991) that controlled environment favoured the destruction of fungal spores both in areas where airborne contamination can occur and in areas of previous fungal damage, while taking observations in the Kremlin's armory collection in Moscow. Most of the problems start with high humidity. A temperature range between 18°C and 25°C and relative humidity (RH) between 45% and 65% is suggested to avoid fungal attack (Plenderleith and Werner 1979).

The use of air-conditioners should be promoted in various museums to maintain the required temperature and RH.

Treatment of display materials should be done by chemicals or by fumigation methods to avoid the damage of objects & further spread of the fungal inoculum to nearby objects.

Use of antifungal chemicals like thymol, naphthalene, paradichlorobenzene, camphor, saffrol etc. should be frequently done in storage cabinets and in showcases to avoid fungal growth.

Periodical inspection of galleries and stores should be done and fungal growth if any detected should be confirmed by isolation. Proper treatment should be done for these objects. Periodical spray of fungicides should be practiced. Vaccum cleaning should be done regularly. It is recommended to make the showcases air-tight to avoid fluctuation of the humidity and temperature.It is also desirable not to open the showcases frequently during monsoon season.

In libraries the excess of humidity and wetting of the valuable books should be avoided. Rare and valuable books should be preserved by thymolated paper.

7.2. Remedial Measures :

It is said that "Prevention is better than cure", but if the microbes have already established themselves in museum objects, we have to adopt certain curative measures. There is an enormous variety of fungicides available in the market but most of them are not suitable for museum purpose. Some may be ineffective to a particular organism. Some are harmful to users and still others can not be recommended as microbes develop immunity against them.

Nair (1974) stated that fungi vary enormously in their response to toxic chemicals. It was found that some fungi are unable to tolerate surprisingly high concentration of certain fungicides which will inhibit other types of fungi at quite low concentration. A fungicide may be obtained from different sources like creosote oil from coaltar; formalin and mercuric chloride from chemicals; Bavistin and Calixin from organic chemicals and cashewnut as well as neem oil as natural plant products.

Considering the past research trends, the following fungicides chemicals/natural plant products have been studied for their efficacy against the four fungi named *C. verruculosus, P. citrinum, A. ustus* and *C. cymbiforme*. Results are summarised in tables 41 to 44.

Fungicides Tested :

i. Bavistin 50% W.P.

ii. Calixin 20% E.C.

Chemicals Tested :

i. Thymol.

ii. Paradichlorobenzene.

iii. Mercuric chloride.
iv. Formalin.
v. Pentachlorophenol.
vi. Creosote oil.
vii. Naphthalene.
viii. Paranitrophenol.

Natural Plant Products Tested :

i. Cashewnut oil.
ii. Neem oil.
iii. Eucalyptus oil.
iv. Godavaj Powder (Sweet flag).

Aqueous Leaf Extracts Tested :

i. Neem: *Azadirachta indica* A. Juss.
ii. Eucalyptus : *Eucalyptus occidentalis* Skeels.
iii. Lantana : *Lantana camara* Linn.
iv. Lemon : *Citrus medica* Linn.
v. Bel : *Aegle marmelos* Correa.
vi. Phudina : *Mentha arvensis* Linn.

7.2 a. Fungicides :

Bavistin is a broad spectrum systemic fungicide manufactured and formulated by BASF India Ltd., Bombay. Chemically it is Methyl 2-benzimidazole carbamate. Of the 2 fungicides tried Bavistin 50% W.P. was more effective giving 100% control of the three organisms at 0.02% concentration. Bavistin could cause 12.5% inhibition of *C. verruculosus* at 0.02% concentration. Use of Bavistin is earlier recommended to control *Botryodiplodia theobromae* (Tandon 1967) and *Phomopsis sapotae* (Lal and Rai 1980).

Calixin is a sophisticated systemic fungicide based on tridemorph which is chemically N-tridecyl-2, 6-dimethyl morpholine. Calixin caused 100% inhibition of *C. verruculosus* and *P. citrinum* at 0.02% concentration. At this concentration *A. ustus* and *C. cymbiforme* were inhibited 33.3% and 70.0% respectively. At 0.03% concentration it could inhibit 100% growth in case of *C. cymbiforme,* whereas, 100% growth inhibition of *A. ustus* was recorded at 0.05% concentration. Calixin was also effective against *Chaetosphaeropsis truncata* var. *indica* (Srivastava *et al.,* 1988) and *Phomopsis gorakhpurensis* (Arya, 1994).

Long back Lilly and Barnett (1951) opined that there is no universal fungicide. Edgington *et al.* (1980) reported that out of 129 fungicides used to control different fungi, 44 were systemic agents. A systemic fungicide possesses therapeutic action unlike protectants, they enter into the system e.g. they can be spread into whole plant through a root or a leaf. They are very selective in nature. In museological studies use of such chemicals is not practiced due to their selective nature.

Both these fungicides can be recommended for control of fungal growth as they produce no adverse effect on museum materials.

7.2 b. Chemical Control :

In the present study eight different chemicals have been tested.

i. *Thymol* : $(CH_3)_2$ CHC_6H_3 (CH_3) OH

Chemically thymol is known as a Isopropyl-meta-cresol. It is frequently used as fumigant. It is a potential fungicide and disinfectant. Workers like Strang and Dawson (1991) reported a number of problems associated with the use of thymol. It causes softening of natural gums and resins, softening of paints and varnishes, yellowing of paper and health problems following prolonged exposure.

The four fungi were grown in medium containing different concentrations of thymol as recommended by Nene and Thapliyal (1979). When thymol (0.01%) was mixed in the medium, *C. verruculosus* failed to grow. In rest of the cases, however, some growth was observed. At a higher concentration (0.02%) none of the four fungi could grow.

"Surface application or spraying with 5% thymol on dried plant materials is found to inhibit fungus growth" (Nair 1991).

Flute-a musical instrument - affected with *C. verruculosus* was treated with 5% thymol by fumigation and subsequently sprayed with alcoholic solution. Five per cent thymol checked the growth of the fungus completely (Shah *et al.* 1992-93).

A wooden plaque affected with *A. niger* was also treated with a 5% thymol by fumigation and after that by spraying an alcoholic solution of 5% thymol. The treatment was followed by a thin coating of picture varnish (Camlin). Reattack of the fungus has not been observed since then (Shah 1990).

ii. *Paradichlorobenzene* : $C_6H_4Cl_2$

It is a white crystalline, volatile substance having penetrating odour. It is toxic by injestion and generally used as a moth repellent. It is also a general insecticide and germicide.

Paradichlorobenzene (PDB) was tested for its efficacy in various concentrations. Table 42 dipicts the results of PDB at five different concentrations. At 0.15% concentration this compound could control 66.6% *C. verruculosus,* 55.5% *P. citrinum,* 62.5% *A. ustus* and 40.0% *C. cymbiforme.* Complete inhibition i.e. 100% control was obtained at 0.25% in all the test organisms. Plenderleith (1957) reported that 1 kg. crystals of PDB are sufficient for one cubic metre of air space for fumigation purpose.

iii. ***Mercuric Chloride*** **:** $HgCl_2$

It is a white crystalline or amorphous powder having no smell and is soluble in water, alcohol, ether, pyridine, glycerol and acetic acid. It is highly toxic by injestion and skin absorption. It is used as disinfectant, fungicide, insecticide and wood preservative. It is used in different surface stabilizing processes due to its antibacterial and anti-fungal property.

In the present study *C. verruculosus* and *P. citrinum* showed no growth at 0.02% concentration while in other two fungi it proved lethal at slightly higher concentration i.e. 0.03%. Gairola (1951) suggested 1% Mercuric chloride as the best fungicide for wood. Agrawal & Dhawan (1958) suggested 0.1% or less concentration of $HgCl_2$ for the protection of wooden material.

Nair (1991) mentioned that dipping of botanical specimens in 1% Mercuric chloride solution (alcoholic or aqueous) prevents the microbial attack. It is a usual practice followed by a plant collector to save the herbaria.

iv. ***Formalin*** **: HCHO**

It is an aqueous 37 to 50% solution of formaldehyde which may contain 15% methyl alcohol. Five to 10% solution of it is used in the wet preservation of Botanical or Zoological specimens. It is a known carcinogen. As it reacts with proteins, it is not recommended for the objects which contain proteins. The role of formalin in seed treatment to make them free from fungi and soil drenching to remove the heavy microbial infestation is well known.

In the present study 0.1% concentration of formalin checked the growth of *C. verruculosus* and *P. citrinum* while in cases of *A. ustus* and *C. cymbiforme* the growth was checked at 0.2% concentration. Nair (1974) proposed 25% formalin for the sterilization of room containing infested paper. Janposri (1991) made use of formalin vapours as a fumigant to control *Aspergillus* and *Penicil-*

lium on paper paintings. Plenderleith (1957) used gaseous formaldehyde for sterilizing room infected with fungus.

v. *Pentachlorophenol* : C_6Cl_5OH

It is colourless substance soluble in alcohol, ether and benzene. It is slightly soluble in water. It is used as fungicide, bactericide, algicide and herbicide. Pentachlorophenol (PCP) is widely applied to wood, textiles, paper, leather and other materials. PCP is highly toxic through inhalation or skin contact (Strang & Dawson 1991). It induces corrosion of iron, degradation of cellulose and discolouration of pigments, paper and wood finishes (Jedrzejewsks 1968).

Another formulation of PCP in the form of Mystox-B, produced by Catomance Limited, U.K. is suggested by Dhawan (1983-84) for the use of leather objects.

In this study a very low concentration i.e. 0.01% could control 100% growth of all the organisms. Nair (1991) recommended the use of 2% PCP to inhibit the growth of fungi. Dhawan (1983-84) suggested the use of 1 to 3% of pentachlorophenol to control the fungal growth. Shah *et al.*, (1991) reported 70-90% inhibition in the growth of *Aspergillus niger, A. sydowii, Penicillium funiculosum* and *Syncephalastrum racemosum* with pentachlorophenol. They also stated that pentachlorophenol is very effective in controlling fungal infection in leather, plastics and paint manufacturing.

vi. *Creosote oil*

Creosote oil is yellowish to dark green-brown in colour derived from the fractional distillation of coaltar. It is used as a wood preservative, disinfectant and fungicide. It is toxic by inhalation.

Creosote oil was mixed in different concentrations with the P.D.A. medium. 0.01% concentration of creosote oil failed to produce any growth in all the four fungi.

vii. *Naphthalene* : $C_{10}H_8$

It is white crystalline, volatile flakes and soluable in benzene, absolute alcohol and ether. It sublimes at room temperature. It is toxic by inhalation and used as a moth repellent and fungicide.It is one of the most common insecticides. The volatile nature of this compound makes it a common antimicrobial compound in museum display cabinets.

It gave 100% control of all the four organisms at 0.02% concentration.

viii. *P-nitrophenol* : C_6H_4OHNO

It is yellowish in colour and monoclinic prismatic crystals, soluble in hot water, alcohol and ether. It turns to strong yellow colour when the pH is above 7.5 and can cause staining of alkaline paper and leather object (Calnan 1985).

Like pentachlorophenol and creosote oil, paranitrophenol (PNP) also controlled *C. verruculosus* and *P. citrinum* at 0.01% concentration. The other two fungi i.e. *A. ustus* and *C. cymbiforme* showed no growth at 0.02% concentration.

Plenderleith (1957) suggested the use of 0.35% PNP in water or alcohol for the protection of leather objects. Mishra & Dhawan (1987) have also suggested that PNP is very much favoured for the protection of leather objects,usually it is used as a spray solution prepared (very low concentration) in Paraffin distillate.

7.2 c. Natural Plant Products as Fungicides :

Highly toxic nature of various fungicides and carcinogenic, pollutive and non-degradable character of these, demands for the use of natural products Botanical pesticides like oils and plant extracts. In the present study effect of cashewnut oil, neem oil, Eucalyptus oil and Godavaj powder was observed on four fungi.

i. Cashew-nut oil :

It is obtained from *Anacardium occidentale* L. of family Anacardiaceae. The plant was introduced in India by Portugues about 400 years ago. The fruit has a kidney shaped nut. Between the two walls of shells, there is a brown oil. It has an extremely blistering effect on the skin. This oil is called the cashew nut oil. The chief constituents of the oil in the shells are Anacardiac acid ($C_{22}H_{32}O_3$), orthohydroxy benzoic acid and cardol ($C_{32}H_{52}O_4$) a phenol. Earlier this oil at 2% concentration has been used for preservation of wooden objects (Shah 1990). He recommended this oil as an insecticide. Ten per cent concentration of cashewnut oil could control 63.6% growth of *C. verruculosus* and 60% growth of *P. citrinum* while only 50% control was achieved in case of *A. ustus* and 45.4% in case of *C. cymbiforme*.

ii. Neem Oil :

Use of neem is a common feature by the people of Indian subcontinent. The twigs of margosa or neem tree (*Azadirachta indica*, A. Juss, family Meliaceae) are used to brush the teeth and its leaves for bathing purposes. The leaves are frequently used for preserving books, textile and grains from insect attack. The neem oil seems to have tremendous potential against the insect pests. The neem oil is extracted from the seeds. It has two crystalline bitter principles- Nimbin and Nimbinin and an amorphous bitter compound Nimbidin. In the bark there is another compound known as Azadirachtin.

In the International Rice Reasearch Institute, Philippines, extracts from powdered leaves and seeds of the neem tree have been tested against a variety of pests. According to a recent newspaper report, different derivatives of neem have been found to repel 123 species of insects including pest of stored grain (Bisht 1991). Use of neem oil in conservation of museum materials has been suggested by Dutta (1985-87). Later on Sharma (1992) suggested use of neem oil alongwith other plants as modern herbal pesticide.

In the present study it was observed that 5% neem oil could control 33.3% growth of *C. verruculosus*, 18.0% growth of *P. citrinum* and 22.2% growth of *A. ustus* and *C. cymbiforme*. Application of neem oil in controlling *Phomopsis psidii* and *P. viticola* has been suggested by Arya (1988-b).

iii. Eucalyptus oil :

Eucalyptus occidentalis Skeels, a member of family Myrtaceae yields volatile eucalyptus oil. The oil is present in the leaves which imparts an aromatic smell. It is used extensively in different medicines. An effort was made to assess its potential as a probable fungicide.

In the present study more than 60% control of *C. verruculosus, P. citrinum* and *A. ustus* was achieved at 0.5% concentration. However, it was only 50% for *C. cymbiforme* at this concentration.

iv. Godavaj Powder (Sweet Flag) :

It is commercially obtained from *Acorus calamus* L. of the family Araceae. It is a small aquatic or marsh herb with fleshy, tuberous, creeping, root-stock and is aromatic in nature.

It is cultivated for its tuberous fleshy rhizomes and used for external application on swelling together with *Curcuma aromatica*. It is also used in *dyspepsia* and *diarrhoea* (Sutaria 1958). "Dry rhizomes of this herb have a mellow odour and contain 1.5 to 3.5% of a yellow, aromatic volatile oil". It is likely that the insecticidal properties of sweet flag are due to the presence of this volatile oil (Agrawal 1984). The rhizomes of *A. calamus* are used medicinally as carminative, stimulant and as a tonic. An essential oil obtained from the rhizomes is used in perfumery. It is commonly found in Kashmir, Manipur and Mysore (Pandey 1969-70).

Godavaj powder at 0.2% concentration gave 36.3% control of *C. verruculosus*, 41.6% of *P. citrinum*, 66.6% of *A. ustus* and 77.7% of *C. cymbiforme*. This compound completely checked the growth of all the four organisms under study at 0.5% concentration.

This powder is used by many libraries in India as an insect repellent. In the present study, this powder was tried as a fungicide by poisoned food technique (Nene and Thapliyal 1979). It showed good fungicidal property. Hence, its use as a museum fungicide is recommended.

7.2 d. II. Aqueous leaf extracts of some plants :

Search of antimicrobial substances in the tissues of higher plants has been initiated recently (Appleton and Tansey 1975; Mishra and Dixit 1976; Arya 1988 a & b). A number of plants have been reported to possess antifungal substances in their leaves (Sekhawat and Prasad, 1971; Khanna and Chandra, 1972; Dixit *et al.*, 1983; Arya and Mathew 1990).

Fiorentin and Bariioni (1991) observed 80% reduction in the growth of 4 *Aspergillus* spp. against grape fruit seed extract. Antifungal nature of garlic bulb is reported by Srivastava and Singh (1982) and Upadhyay and Gupta (1990). The volatile fraction from the leaf of garlic (*Allium sativum*) inhibited the spore germination of four Deuteromycetous fungi (Mishra and Dixit 1979). Arya *et al.* (1995) reported sulphur containing compound Allicin in the leaves of *Allium* and *Bignonia* which inhibited growth of *A. niger* and *F. pallidoroseum*. *Eucalyptus* leaves may be effective due to the presence of volatile compounds and Cineole in them (Arya 1988b). It has been observed that the activity of extracts may be lost due to autoclaving (Lokesha *et al.* 1986).

Neem leaves are used as pesticide since long. We can find their use in our daily life. These are mixed with clothes, paper and grains to keep them free from insect and fungal pests. Two concentrations were tried, at 100% concentration 50% control was observed in *A. ustus* followed by 40% in *C. verruculosus* and *C. cymbiforme* and 30% in *P. citrinum*. Earlier this plant has been tried by a number of workers. Arya (1988b) recorded 82.3% spore inhibition of *Phomopsis viticola* in 75% concentrated leaf extract. This extract caused only 10% germination of spores of *Pestalotia psidii* (Pandey *et al.*, 1983).

Extract from leaves of *Eucalyptus* was slightly less effective than neem.

50% extract was ineffective for *P. citrinum* and *C. cymbiforme*. 100% extract controlled 30% growth of *C. verruculosus*, 25% of *A. ustus*, 20% of *P. citrinum* and 10% of *C. cymbiforme*. Earlier Arya (1988-a) found that 75% extract of *E. occidentalis* inhibited 83.33 and 78.82% spore germination of *P. psidii* and *P. viticola* respectively.

Lantana a member of Verbenaceae, was another effective plant. It is a common weed but can be grown for its spectacular small flowers in a bunch. 100% leaf extract of this shrub could control 50% or more than 50% growth of *P. citrinum, A. ustus* and *C. cymbiforme*. Control was 33.3% only in case of *C. verruculosus*. The extract of *L. camara* exhibited high antifeedant and repellent activity against furniture beetle *Sinoxylon sudanicum* (Urs and Reddy 1991). Shri Agrochemicals Nandesari, Baroda has produced a fungicide named Ovis from the leaves of *L. camara* (Cortellini and Chinnella 1992).

Leaves of three more plants *Citrus medica*, (Lemon), *Mentha arvensis* (Phudina) and *Aegle marmelos* (Bel) were tried at 50 and 100% concentrations. *C. medica* L. is an evergreen shrub or small tree belonging to the family Rutaceae. The leaves are unifoliate, gland-dotted and contain volatile oil like neroli and petitgrain-*oils*. *M. arvensis* L. or phudina is a member of family Lamiaceae. The plant has characteristic aromatic leaves, square stems and flowers with bilabiate corolla. It is a perennial herb. Green leaves contain volatile oil having menthol. This may be the antifungal compound present in the leaves.

The 'Bel' (*Aegle marmelos* Corr.) is an important medicinal tree of Indian origin. It belongs to the family Rutaceae. Almost all the parts of this tree are used to cure various diseases. The leaves contain an alkaloid aegelinine and an essential oil cineole having antifungal property. Of the three plants leaves of Bel were more effective followed by *M. arvensis*. *C. medica*, however, failed to control growth of any fungus. Earlier Arya and Mathew (1990) reported 17.6% inhibition of *A. niger* spores by 75% leaf extract of *A. marmelos*.

Table - 41 : Effect of two fungicides on the percentage control of four test organisms

Fungicides	conc. %	*C. verruculosus* Colony diameter* Control	Treat-ment	% Control	*P. citrinum* Colony diameter* Control	Treat-ment	% Control	*A. ustus* Colony diameter* Control	Treat-ment	% Control	*C. cymbiforme* Colony diameter* Control	Treat-ment	% Control
1. Bavistin	0.01	-	-	-	4.5	0.0	100.0	4.5	0.0	100.0	5.0	0.0	100.0
50% W.P.	0.02	4.0	3.5	12.5	-	-	-	-	-	-	-	-	-
	4.00	4.0	1.0	75.0	-	-	-	-	-	-	-	-	-
	8.00	4.0	0.0	100.0	-	-	-	-	-	-	-	-	-
2. Calixin	0.02	4.0	0.0	100.0	4.0	0.0	100.0	4.5	3.0	33.3	5.0	1.5	70.0
80% E.C.	0.03	-	-	-	-	-	-	4.5	2.5	44.0	5.0	0.0	100.0
	0.04	-	-	-	-	-	-	4.5	1.5	66.6	-	-	-
	0.05	-	-	-	-	-	-	4.5	0.0	100.0	-	-	-

* = Average diameter of 3 fungal colonies measured in cm.

- = Shows treatments were not tried.

Table - 42 : Effect of different chemicals on the precentage control of four test organisms.

Chemicals	conc.	*C. verruculosus*			*P. citrinum*			*A. ustus*			*C. cymbiforme*		
		Colony diameter*			Colony diameter*			Colony diameter*			Colony diameter*		
	%	Control	Treatment	% Control	Control	Treatment	% Control	Control	Treatment	% Control	Control	Treatment	% Control
1. Thymol	0.01	4.0	0.0	100.0	3.5	1.0	71.4	7.0	4.0	42.8	6.0	2.8	53.3
	0.02	-	-	-	3.5	0.0	100.0	7.0	0.0	100.0	6.0	0.0	100.0
2. Paradichlo robenzene	0.01	6.0	3.5	41.6	4.5	3.0	33.3	4.0	2.0	50.0	5.0	4.0	20.0
	0.05	6.0	3.0	50.0	4.5	3.0	33.3	4.0	2.0	50.0	5.0	3.5	30.0
	0.10	6.0	2.0	66.6	4.5	3.0	33.3	4.0	2.0	50.0	5.0	3.5	30.0
	0.15	6.0	2.0	66.6	4.5	2.0	55.5	4.0	1.5	62.5	5.0	3.0	40.0
	0.25	6.0	0.0	100.0	4.5	0.0	100.0	4.0	0.0	100.0	5.0	0.0	100.0
3. Mercuric chloride	0.01	6.0	2.5	60.0	4.5	2.0	55.5	6.5	2.5	62.0	5.0	1.5	70.0
	0.02	6.0	0.0	100.0	4.5	0.0	100.0	6.5	1.5	77.0	5.0	0.5	90.0
	0.03	-	-	-	-	-	-	6.5	0.0	100.0	5.0	0.0	100.0
4. Formalin	0.01	4.0	2.0	50.0	4.5	3.0	33.3	3.5	2.5	29.0	5.0	3.0	40.0
	0.05	4.0	1.5	62.5	4.5	1.5	66.6	3.5	2.0	43.0	5.0	2.0	60.0
	0.10	4.0	0.0	100.0	4.5	0.0	100.0	3.5	0.5	86.0	5.0	0.5	90.0
	0.20	-	-	-	-	-	-	3.5	0.0	100.0	5.0	0.0	100.0
5. Pentach- lorophenol	0.01	6.0	0.0	100.0	5.0	0.0	100.0	4.5	0.0	100.0	5.0	0.0	100.0
6. Creosote oil	0.01	6.0	0.0	100.0	4.5	0.0	100.0	4.5	0.0	100.0	5.0	0.0	100.0
7. Naphthalene	0.02	6.0	0.0	100.0	4.5	0.0	100.0	4.5	0.0	100.0	5.0	0.0	100.0
8. P. nitro- phenol	0.01	4.5	0.0	100.0	5.0	0.0	100.0	4.5	1.5	66.6	5.0	1.5	60.5
	0.02	-	-	-	-	-	-	4.5	0.0	100.0	5.0	0.0	100.0

* = Average diameter of 3 fungal colonies measured in cm.

- = Shows treatments were not tried.

Table - 43 : Effect of different oils and plant products on the percentage control of four test organisms.

Oils and Plant Products	conc. %	*C. verruculosus*			*P. citrinum*			*A. ustus*			*C. cymbiforme*		
		Colony diameter*			Colony diameter*			Colony diameter*			Colony diameter*		
		Control	Treat-ment	% Control	Control	Treat-ment	% Control	Control	Treat-ment	% Control	Control	Treat-ment	% Control
1. Cashew nut oil	1.0	5.5	4.0	27.2	5.0	3.0	40.0	5.0	3.5	30.0	5.5	3.5	36.3
	2.0	5.5	3.5	36.3	5.0	2.5	50.0	5.0	3.5	30.0	5.5	3.5	36.3
	4.0	5.5	3.0	45.4	5.0	2.5	50.0	5.0	2.5	50.0	5.5	3.5	36.3
	10.0	5.5	2.0	63.6	5.0	2.0	60.0	5.0	2.5	50.0	5.5	3.0	45.4
2. Neem oil	2.0	6.0	5.0	16.6	5.5	5.0	9.0	4.5	3.5	22.2	4.5	3.5	22.2
	5.0	6.0	4.0	33.3	5.5	4.5	18.0	4.5	3.5	22.2	4.5	3.5	22.2
3. Eucalyptus oil	0.2	6.5	3.5	46.0	5.5	3.2	45.4	5.5	3.5	36.3	5.0	4.0	20.0
	0.5	6.5	2.5	62.5	5.5	2.0	63.6	5.5	2.0	63.6	5.0	2.5	50.0
4. Godavaj Powder	0.2	5.5	3.5	36.3	6.0	3.5	41.6	4.5	1.5	66.6	4.5	1.0	77.7
	0.3	5.5	2.0	63.6	6.0	2.5	58.3	4.5	1.0	77.7	4.5	0.5	88.8
	0.5	5.5	0.0	100.0	6.0	0.0	100.0	4.5	0.0	100.0	4.5	0.0	100.0

* = Average diameter of 3 fungal colonies measured in cm.

Table - 44 : Effect of aqueous leaf extracts of some plants on the percentage control of four test organisms.

Leaf Extracts	conc.	*C. verruculosus*			*F. citrinum*			*A. ustus*			*C. cymbiforme*		
		Colony diameter*			Colony diameter*			Colony diameter*			Colony diameter*		
	%	Control	Treat-ment	% Control	Control	Treat-ment	% Control	Control	Treat-ment	% Control	Control	Treat-ment	% Control
1. Neem	100.0	5.0	3.0	40.0	5.0	3.5	30.0	4.0	2.0	50.0	5.0	3.0	40.0
	50.0	5.0	4.0	20.0	5.0	4.5	10.0	4.0	3.0	25.0	5.0	4.0	20.0
2. Eucalyptus	100.0	5.0	3.5	30.0	5.0	4.0	20.0	4.0	3.0	25.0	5.0	4.5	10.0
	50.0	5.0	4.0	20.0	5.0	5.0	0.0	4.0	3.5	12.5	5.0	5.0	0.0
3. Lantana	100.0	4.5	3.0	33.3	4.0	2.0	50.0	4.5	2.0	55.5	5.0	2.5	50.0
	50.0	4.5	3.5	22.2	4.0	3.0	25.0	4.5	2.5	44.4	5.0	3.5	30.0
4. Lemon	100.0	4.5	4.5	0.0	5.0	5.0	0.0	4.0	3.5	12.5	4.5	4.5	0.0
	50.0	4.5	4.5	0.0	5.0	5.0	0.0	4.0	4.0	0.0	4.5	4.5	0.0
5. Bel	100.0	5.5	4.5	18.1	6.0	5.5	8.3	5.5	3.0	45.4	5.0	3.5	30.0
	50.0	5.5	4.5	18.5	6.0	5.5	8.3	5.5	3.5	36.3	5.0	4.5	10.0
6. Phudina	100.0	5.5	5.5	0.0	6.0	5.0	16.6	5.5	3.5	36.3	5.5	5.5	0.0
	50.0	5.5	5.5	0.0	6.0	5.5	8.3	5.5	4.5	18.1	5.5	5.5	0.0

* = Average diameter of 3 fungal colonies measured in cm.

Chapter—8

Conclusions

The results obtained in the present investigation have been discussed fully at appropriate places in different chapters. In the present chapter, however, efforts have been made to correlate the results of various experiments and certain conclusions have been drawn wherever possible.

Conservation is necessary when deterioration occurs or is likely to occur. Among the agencies of deterioration biological agencies are the most devastating on art objects. All types of art material – monuments, stone sculptures, paintings and manuscripts are vulnerable to damage by biological factors and therefore, it becomes an important issue to take care of cultural property. Fungus and other micro-organisms are major contributors to the deterioration of cultural property in the tropical conditions.

Present *in vitro* studies provide information on the preference of nutrients utilized by *C. verruculosus, P. citrinum, A. ustus* and *C. cymbiforme* and factors governing their growth and sporulation. Possible control measures are suggested at the end.

Among the seven culture media tried, Richard's medium showed the best growth of all the fungi under present study except *A. ustus* which showed the best growth in Czapek's medium.

The organisms under study tolerated a wide range of initial pH i.e. 3 to 10, which shows their capability to survive under fluctuating pH of the different museum materials.

Amongst physical environmental factors, it was found that the biodeteriogens thrived best at 25^{0}C, which coinsides fairly closely with the range that generally prevails in tropics.

Carbohydrates being the chief metabolic product, have special importance in the nutrition of the fungi. The availability of the carbon source in nature may be both, in simple as well as in complex state. All the four organisms were capable of producing satisfactory growth on above compounds. Glucose and fructose produced best growth in *P. citrinum*. Sucrose a common oligosaccharide was the best source of carbon for *C. cymbiforme*. Two other fungi

C. verruculosus and *A. ustus* produced best growth in carboxy methyl cellulose. Present organisms differed considerably in their response towards the various sources of carbon. Conversion of higher sugars into lower ones before their utilization indicated the utility of monosaccharides in the nutrition of fungi.

Chromatography is a recent technique to detect compounds even in small quantity. Utilization of sugars-separately and in mixtures-was studied chromatographically. In case of *C. verruculosus* sucrose was present upto 4 days, after which it was hydrolysed. Glucose remained present upto 6 days. Maltose was also utilized through a hydrolytic pathway by the present fungi.

Effect of 9 different sources of nitrogen was studied. Potassium nitrate produced best growth in all the four fungi. Nitrates of cobalt, copper and thiourea were toxic as no growth was recorded in these cases. A medium devoid of any nitrogen source failed to support the growth in any fungus. Lakshmikant and Mathur (1991) concluded "On the basis of nutrition requirement of cellulolytic fungi, a recommendation can be made to make papers free from nitrogen for various use, because nitrogen content in the cellulosic materials is directly responsible for growth and activity of cellulolytic fungi." The compounds like cobalt and copper nitrates which were inhibitory can be tried as fungicides.

Sulphur is an important constituent of protein. It plays a significant role in the metabolic activities of fungi. Effect of 10 different sources of sulphur was studied on four fungi. Except *A. ustus* all the fungi supported best growth in magnesium sulphate. Ferrous sulphate was the best source of sulphur for *A. ustus.*

Growth and sporulation of 4 fungi on 7 different sources of phosphorus were studied. Potassium hydrogen phosphate, potassium dihydrogen phosphate and sodium phosphate induced good growth of all the four fungi. Variations existed amongst the different biodeteriogens for the comparative choice of these substances.

To get a correct picture of fungi present in the museum, a survey of indoor aeromycoflora was done. In this study 22 different fungi belonging to 12 genera were recorded. Many of these fungi have been reported earlier causing serious damage to cultural property. Some of these are plant pathogens and human allergens. A regular monitoring of these fungal spores is suggested. The presence of various fungal inocula in the indoor atmosphere of museums and libraries clearly warrants immediate protective measures like reduction of humidity levels and maintenance of temperature in the galleries to avoid fresh attack on museum materials.

Wood is extensively used in museum collections particularly in the ethnographical and folk exhibits. Museum furnitures, storage cabinets and showcases are made-up of wood. Thus it is a multipurpose material. Percentage loss, final pH and sporulation of 5 different wood chips by four organisms was studied for 25 days. Five to 17.6% of reduction in wood weight was observed in this experiment. Wood is an ideal carbon source but is highly polymeric and thus insoluble. Extra cellular fungal enzymes depolymerize the different cell wall constituents into water soluble fragments. Besides the chemical nature of different woods their texture or presence of wood elements in various forms are responsible for variation in their biodeterioration.

Paper is one of the best writing materials, earlier palm leaves and bark of the trees were used for writing manuscripts. It is a very complex hygroscopic system within homogeneous network of pores and capillaries. Paper is destroyed commonly by *Aspergillus, Chaetomium, Nigrospora* etc. Percentage loss of 8 different varieties of paper by four fungi was studied. Two organisms had a choice for card board. Rate of biodegradation was different in various types of papers. Paper deterioration is determined by its properties like nutrient substances present, composition of fibre, sizing fillers, ash content, metal inclusions, pH of paper and weight of paper/m^2 etc. Since all these parametres were different for 8 papers, this difference in percentage loss is recorded.

Preventive and Remedial measures are suggested. Importance of good house keeping habit and proper sanitation, control of environmental condition, treatment of display material, periodical inspection and precaution in libraries is discussed in detail.

If a microbe settles on a museum object remedial measures are not the only answer. Chemical treatment with two fungicides and 8 chemicals is tried. Bavistin an organic systemic fungicide was superior than Calixin. Of the 8 antifungal compounds tried, pentachlorophenol and creosote oil were most effective followed by thymol, naphthalene and p-nitrophenol. *C. verruculosus* was completely inhibited by 0.01% concentration of thymol, pentachlorophenol, and p-nitrophenol. For *P. citrinum* pentachlorophenol, creosote oil, nephthalene and p-nitrophenol were most effective antifungal chemicals at 0.01% concentration. *A. ustus* can be effectively controlled by 0.01% pentachlorophenol and creosote oil. Complete inhibition of *C. cymbiforme,* fourth test fungus, was recorded at 0.01% concentration of pentachlorophenol and creosote oil.

Agrawal (1991) is of the opinion that before suggesting any fungicide we should check;

a. Its toxic effect on human beings,

b. Long and short term effect of such chemicals on museum objects,

c. Time of exposure of museum materials with the fumigants.

Most of these fungicides and chemicals show carcinogenicity, teratogenicity, pollutive effects and residual toxicity and thus are not eco-friendly. Arsenic and mercury containing fungicides are health hazards, formalin is unfit for use on museum materials. According to Strang and Dawson (1991) paranitrophenol, parachlorometacresol and thymol are not recommended for use in Canada. Certain natural plant products, oils and leaf extracts were tried as they are less toxic and have easy biodegradability. Aqueous leaf extracts of neem, *Eucalyptus, Lantana* and bel were effective against four biodeteriogens. Possibilities should be explored to use these plants as such or in different formulations against such fungal biodeteriogens.

Chapter–9

Summary

Cultural heritage is perishable. At the same time it is the prime duty of every person to pass this valuable cultural heritage to the posterity. Damage to cultural property may be due to different reasons. Biological agency is one of them. Insects, bacteria and fungi are few of such destructive agents. It is said that correct strategy cannot be evolved unless the biology, life history and ecology of the pest are known in detail. Detailed structure and biochemical requirements are investigated, efforts have been made to suggest some chemicals/fungicides and natural plant products which are non pollutive and quite safe to use.

Following investigations were undertaken :

1. A detailed survey of museum objects was undertaken and fungi causing damage were isolated. Moreover, indoor aeromycoflora of different institutions revealed presence of several fungi in the atmosphere. Some of the fungi found present in the atmosphere were also observed causing damage to various organic objects displayed in the museum gallery. Among these fungi *Cochliobolus verruculosus* (Tsuda and Ueyama) Sivan, *Penicillium citrinum* Thom., *Aspergillus ustus* (Bainier) Thom. & Church and *Chaeotomium cymbiforme* Lodha were selected for detailed study. These fungi were isolated from wood and paper materials. These were purified and identified. Cultures were maintained on potato dextrose and agar medium. It is interesting to note that *C. verruculosus* and *C. cymbiforme* are new reports from museum materials.
2. Morphology of the four fungi was studied with the help of light microscope.
3. Organisms were grown on seven different culture media. The maximum growth was obtained on Richard's medium. Since the sporulation was good in modified Asthana and Hawker's medium 'A' and as it was easy to handle and manipulate according to the need of experiments, this medium was selected for further cultural studies.

4. All the four fungi tolerated a wide range of initial pH (3.0 to 10.0). *C. verruculosus* and *C. cymbiforme* failed to grow at pH 2.0. It was observed that the increase in pH after neutral showed decrease in mycelial dry weight. In most of the cases growth was minimum at pH 10. The most suitable pH for good growth and excellent sporulation for all the four fungi was 5.5. Thus for all subsiquent cultural studies, pH of the medium was adjusted to 5.5.
5. The maximum growth and good sporulation was recorded at 25⁰C. All the organisms studied were failed to grow at low temperature of 5⁰C except *C. verruculosus. C. verruculosus* and *P. citrinum* failed to grow at 45⁰C while *A. ustus* failed to grow at 55⁰C and *C. cymbiforme* at 50⁰C.
6. The influence of 9 different carbon sources on growth and sporulation of these fungi was studied. All the four organisms were capable of growing on various sources of carbon except oxalic acid but the amount of growth varied in individual carbon sources. All these organisms failed to grow in a medium devoid of any carbon source.
7. Daily utilization of mono-saccharides, disaccharides and polysaccharides was studied chromatographically in culture medium by four fungi. *P. citrinum* and *A. ustus* utilized glucose in 4 days, *C. verruculosus* in 12 days and *C cymbiforme* failed to utilize it upto 15 days. Fructose was a favourable source for *C. verruculosus, P. citrinum* and *A. ustus.* Of the three disaccharides sucrose was preferencially utilized by all the fungi except *C. verruculosus.*
8. Nitrogen is most essential element for the growth of fungi. The effect of 9 different nitrogen compounds was studied. Thiourea, copper nitrate and cobalt nitrate proved lethal to all the fungi while other nitrogen compounds showed moderate growth in all the organisms. Maximum growth and good to excellent sporulation was recorded in potassium nitrate. Hence it was the best source of nitrogen for these fungi.
9. The effect of 10 different sources of sulphur on growth and sporulation was investigated. out of these, thiourea and copper sulphate were found toxic. Magnesium sulphate supported good growth in all the fungi. Growth was almost negligible in the medium devoid of any sulphur source.

10. All the organisms were tested for seven different phosphorus sources. Potassium dihydrogen phosphate, potassium hydrogen phosphate and sodium phosphate induced good growth in all the cases while sporulation was found good in potassium dihydrogen phosphate. However, wide variations among the four fungi regarding their choice for different phosphorus compounds have been observed. It was also found that phosphorus was an essential element for the growth of fungi.
11. Indoor aeromycological study was conducted in 4 different institutions of Baroda. This study revealed the presence of fungal spores of different genera. Among these some were responsible for deterioration of museum objects in various forms while some were potential allergens to human beings.
12. Effect of artificial deterioration of different woods by four fungi was studied. Results revealed that deodar wood was more susceptible to these organisms and percentage loss occurred in this wood was comparatively more. Minimum percentage loss was observed in neem wood.
13. In ancient time, primitive people used various materials to write upon but the invention of paper provided a better alternative to all those materials. Biodeterioration of 8 different types of paper was studied. Results showed the maximum degradation in Japanese tissue paper by *C. verruculosus* and *A. ustus* whereas in card board by *P. citrinum* and in brown paper by *C. cymbiforme*. Thus it can be concluded that choice of paper as a food may differ from fungus to fungus.
14. Control of fungus is our ultimate aim. When any museum object harbours any fungus, its extreme need is to kill the fungus, in order to check the repeated fungal attack and to avoid secondary infection. To find out a suitable fungicide, it is extremely necessary to study the behaviour of the fungi. In the present study various chemicals/fungicides/natural plant products were tried. Emphasis has been made to find out minimum lethal dose of different fungicides for fungi under study.

Certain preventive measures are recommended too to avoid fungal attack on museum materials.

References

Adisa V.A. 1988 : Some nutritional requirements of spoilage molds of Pineapple fruit. *Cryptogam Myco.* 9 (2) : 111 - 116. c.f. Biological Abstracts *86* (10) : 101812.

Agnihotri V.P. 1968 : Effect of nitrogen compounds on sclerotium formation on *Aspergillus niger*. *Can J. Microbiol.* 14, 1253-1258.

Agrawal M.K., Mukerji K.G. & Shivpuri D.N. 1968 : Studies on the allergenic fungal spores of Delhi, India, Metropolitan area. *J. of Allergy,* 44 (4) : 193-203.

Agrawal, O.P. 1975 : Problems of Preservation of Ajanta Wall Paintings. *Con. of Cul. Prop. in India,* 3 : 13-21.

Agrawal, O.P. 1984 : *Conservation of Manuscripts and Paintings of South-East Asia.* Butterworths, London. pp. 299.

Agrawal, O.P. 1989 : *Examination and conservation of wall paintings of Sheesh Mahal, Nagaur,* N.R.L.C. & INTACH, Lucknow, pp. 58.

Agrawal, O.P. 1991 : An overview of studies on biodeterioration of cultural property. In, "*Biodet. of Cul. Prop.*" (Eds. Agrawal, O.P. & Dhawan, S.) MacMillan India Ltd., New Delhi : 3-15.

Agrawal, O.P. 1993 : *Preservation of Art Objects and Library Materials,* National Book Trust, New Delhi, pp. 102.

Agrawal, O.P. 1994 : An overview of Problems of Biodeterioration of Cultural Property in Asia. *Con. of Cul. Prop.* in Asia. *Con. of Cul. Prop.* in India, New Delhi. 27 : 90-107.

Agrawal, O.P. 1995 : Conservation of Tajmahal—The Real Problem. In, Conservation, Preservation & Restorations, Traditions, Trends and Techniques (Eds. Kamalakar, G. & Rao, V.P.), Birla Archaeological & Culural Research Institute, Hyderabad, 1-14.

Agrawal, O.P. & Dhawan, S. 1985 : *Control of Biodeterioration in Museums,* Technical Note : 2, N.R.L.C., Lucknow, pp.16.

Agrawal, O.P., Dhawan, S. & Garg, K.L.	1990 :	*Microbial Deterioration of Wood A. Review*. INTACH (ICI), Lucknow, pp. 66.
Agrawal, O.P. & Pathak, N.	:	*Bio-deterioration of Ethnological objects*. INTACH (ICI), Lucknow, pp. 42. (Year not mentioned).
Ainsworth, G.C., 1966	:	A General purpose classification of Fungi. *Bibi. Syst. Mycol.* No. 1 : 1-4.
Alexopoulos, C.J., 1962	:	Introductory Mycology John Wiley and Sons, New York and London.
Anchabagze, T.T. 1949	:	Materials for the Study of Book Destroying Fungi. *Vest. Gos. Muzeia Gruzii in Dzhanashia* : 14 A : 201-206.
Ansari, N.A., Khan, M.W. & Muheet, A.	1989 :	Nutritional requirements of *Alternaria brassicae for growth and sporulation. Ind. J. of plant pathol.* 7 (2) : 125-135. c.f. *Rev. of pl, path.* 71, 1992 : 3867.
Appleton J.A. & Tansey, M.R.	1975 :	Inhibition of growth of zoopathogenic fungi by garlic extract. *Mycologia* 67 : 882-885.
Arai, H.	1987 :	Microbiological stuides on the conservation of paper and related cultural property (part-2) characteristics of fungi isolated from foxing formation machanism and counter measures, *Science for con servation* 26 : 43-52.
Arya, A.	1982 :	Cultural and pathological studies of certain Fungi imperfecti, D. Phil, Thesis, Allahabad University, pp. 246.
Arya, A.	1988-a:	Control of Phomopsis fruit rots by leaf extracts of certain medicinal plants. In, "*Indigenous Medicinal plants symp.*" (Ed. P. Kaushik) Today & Tomorrow's Printers & Publishers, New Delhi : 41-46.
Arya, A.	1988-b:	Control of Phompsis fruit rots of grapes and guava *Indian Phytopath.* 41 (2) 214-219.
Arya, A.	1990 :	Influence of hydrogen-ion concentration on the utilization of sodium nitrite by four different species of *Phomopsis. Proc. Nat. Acad. Sci., India,* 60 (B), III : 359-361.

Arya, A. 1990-91: Nutritional requirements of two Pathogenic Species of *Phomopsis*. Science Tech. & Medicine J. of M.S. University Baroda, 37-38 (3) : 71-79.

Arya, A. 1991 : Nutritional requirements of two species of *Phomopsis*. *Res. J. of Pl. Environ.* 7 : 15-25.

Arya, A. 1994 : A new leaf blight of *Polyalthia longifolia (Sonn.)*. Thw. and its chemical control. *Res. J. Pl. Env.* 10 : 29-31.

Arya, A. & Mathew, D. 1990 : Control of chiku fruit rot by leaf extracts of certain Medicinal Plants. *Res. J. Pl. Env.* 6 (1) : 31-33.

Arya, A., Chauhan, R. & Arya, C. 1995 : Effect of Allicin and extracts of garlic and *Bignonia* on two fungi. *Ind. J. Mycol. and Pl. Pathol.* 25 : 316-318.

Barkley, F.A. 1968 : Outline classification of Organisms, 2nd ed., Hopkins Press, Providence, Massachusets.

Basu, S.N. & Ghosh, S.N. 1950 : Fungi growing on jute. *J. Sci. Indutr. Res. India,* 9 B, 6 : 151-156.

Baxi, S.J. & Dwivedi, V.P. 1973 : *Modern Museums-organisation & Practice in India.* Abhinav Publications, New Delhi-pp. 208.

Bhargava, S.N. 1962 : Physiological and Pathological studies of some fungi causing storage rots. D. Phil. Thesis, University of Allahabad.

Bhargava, S.N. & Tandon, R.N. 1963 : Vitamin requirements of three Pathogenic fungi. *Path. Micro biol.* 26 : 313-316. c.f. Rev. of *Appl. Mycol.* 43 (8) 1964 : 2222.

Bhatnagar, I.K. 1991 : Problem of Biodeterioration of Kadathus of Karnataka. In, "*Biodet. of Cul. Prop.*" (Eds. Agrawal, O.P. & Dhawan, S.) MacMillan India Ltd. New Delhi. : 83-90.

Bilgrami, K.S. 1956 : Pathological and Physiological studies of some fungi causing leaf spot diseases. D. Phil. Thesis University of Allahabad.

Bilgrami, K.S., Jamaluddin, & Rizwi, M.A. 1979 : *Fungi of India* Part-I. Today & Tommorows, Print & Publ. New Delhi. pp. 467.

Bilgrami, K.S., Jamaluddin, & Rizwi. M.A. 1981 : *Fungi of India* Part-II. Today & Tommorows, Print & Publ. New Delhi. pp. 128.

Bilgrami, K.S. & Tandon, R.N. 1957 : The assimilation of sugars by three pathogenic species of *Phyllosticta. Proc. Nat. Acad. Sci. India.* 27 B : 196-203.

Bilgrami, K.S. & Verma, R.N. 1978 : *Physiology of Fungi.* Vikas Pub. House Pvt. Ltd., New Delhi, pp. 507.

Bisht, A.S. 1991 : Control of Biodeterioration in museums - A study of Problem. In, "*Biodet. of Cul. Prop.*" (Eds. Agrawal, O.P. & Dhawan, S.) MacMillan India Pvt. Ltd., New Delhi : 93-103.

Brefeld, O. 1881 : Botani. Unte. Uber. Schim. Heft. Hein. Schon. Muen : 1-4.

Calnan, C.N. 1985 : *Fungicides used on Leather.* Leather Conservation Centre, London. c.f. *Controlling Museum Fungal Problems* by Strang and Dawson, 1991, C.C.I. Publication, Canada.

Chander, H., Sannabhadi, S.S., Elias, J. and Ranganathan, B. 1977 : Factors effecting lipase production by *Penicillium chrysogenum. J. Food Sci.* 42 : 1677-1682.

Chitaley, S.D. & Bajaj. A. 1973 : Airspora of Nagpur of high altitudes-I. *The Botanique* 4 (1) : 27-34.

Chowdhury, S.R. 1981 : Assimilation of sugar by *Phomopsis vexans* (Sacc. and Syd.) Harter. The incident of soft fruit rot disease of *Solanum melongena. Proc. Nat. Acad. Sci. India* 51 B : 74-84.

Chung, H.S. 1967 : Germination of conidia and Reporduction in *Phoma herbarum* - var- *medicaginis.* Thesis abstract *Medicago - sativa* - D. Medicago - *dzawkhetica* - D. c. f. Abstracts of Mycology 2 (5) : 293 : 1968. *Diss. Abstr.* 38 (2) B: 420.

Cochrane, V.W. 1958 : *Physiology of Fungi.* John Willey & Sons. Inc. New York. pp. 524.

Cohen, B.L. 1973 : The neutral and alkaline proteases of *A. nidulans. J. Gen. Microbiol* 109 : 265-274.

Copeland, H.F. 1956 : The Classification of Lower Organisms. Pacific Books, Polo Alto, California.

Coremans P. 1960 : The Museum Laboratory. In *The organisation of Museums,* Practical Advice UNESCO: 93-118.

Cortellini, W. 1992 : Ovis, Bio efficacy trial Reports. Central Experimental Research Centre, Itlay, not paginated.

Deshpande, M.N., 1959 : Important epigraphical records from the Chaitya Cave, Bhaja. *Lalitkala* 6 : 30-32.

Deverall, B.J. 1965 : The Physical environment for fungal growth, I-temperature. In, "*The Fungi.*" (Eds. Ainsworth, G.C. and Sussman, A.S.) Vol.-1, Academic Press, New York : 543-550.

Devey, C.B. & Papavizas, G.C. 1962 : Growth and sexual reproduction of *Aphanomyces euteiches* as affected by the oxidation state of sulphur. *Amer. J. Botany* 49 (4) : 400-404.

Davies, R.R. & Smith, L.P. 1973 : Aspects of human allergy and fungal spores. *Clini cal Allergy* 3 : 263-267.

Dhawan, S. 1983-84 : Microbial deterioration of leather materials. A lit. review. *Con. of Cul. Prop. in India.* 16-17 : 156-163.

Dhawan, S. 1986 : Microbial deterioration of paper material. A lit, review. NRLC. Lucknow : pp. 18.

Dhawan, S. & Agrawal, O.P. 1986 : Fungal flora of miniature paintings & lithographs. *Int. Biodet. Bull.* 22 (2) : 95-99.

Dhawan, S., Mishra, A., Garg, K. & Pathak, N. 1991 : Laboratory evaluation of ortho phenyl phenol & P. chloro-m-cresol for the control of some Fungal forms of Ajanta wall Paintings, in, "*Biodet. of Cul. Prop.*" (Eds, Agrawal, O.P. and Dhawan, S.) MacMillan India, Ltd., New Delhi : 313-338.

Dixit, S.N., Dube, N.K. & Tripathi N.N. 1983 : Fungitoxic essential oils vis-a-vis disease control. In, "*Recent Advances in plant Pathol.*" (Eds. Husain, A., Singh, K., Singh, B.P. and Agnihotri, V.P.) Print House (India) Lucknow, pp. 248-252.

Dube, H.C. 1990 : *An Introduction to Fungi,* Vikas Publishing House Pvt. Ltd., 608 pp.

Dutta, P.K. 1985-87 : Use of Neem Oil in Conservation, *Con. of Cul. Prop. in India,* 28-30, New Delhi : 98-100.

Feilden, B.M. 1991 : Conservation of Worlds heritage - A personal view. In, "*Recent Trend in Conservation of Art Heritage.*" (Ed. Dhawan, S.) Agam Kala Prakashan, Delhi : 1-20.

Fiorentin, L. & Bariioni Junior 1991 : Inhibition of *Aspergillus flavus* growth by grape fruit seed extract. *A. rguiro Brasileiro de Medicina Verterinariae Zootechnia Brazil.* 43 (3) : 227-239. c. f. *Rev. Pl. Path,* 1992, *71* : 6053.

Foster, R.A. 1986 : The larger provincial museums and art galleries. In, *Manual of curatorship* Butterworths, London : 66-75.

Gairola, T.R. 1951 : Preservation of Wooden Antiquities. *J. Ind. Museums,* 7 : 25-35.

Garg, K.L. & Dhawan, S. 1985-87 : Microbial deterioration of wooden cultural property. *Con. of Cul. Prop. in India.* 18-20 : 74-85.

Garg, K.L. & Mukerji, K.G. 1994 : Deterioration of Wooden Materials by Microorganisms. In, "*Recent Advances in Biodeterioation and Biodegradation.*" Vol. I (Eds. Garg, K.L., Garg, N. and Mukerji, K.G.) Naya Prokash, Calcutta : 429-479.

Ghose, A.K. & Tandon, R.N. 1965 : Utilization of polysaccharides by some anthracnose fungi. *Proc. Nat. Acad. Sci. India.* 35 B : 383-389.

Golubic, S., Friedmann, I., & Schneider, J. 1981 : The Lithobionthic ecological niche, with special reference to microorganism., *J. Sediemtary Petrology* 5 : 475-478.

Gonzaler, H., Silvial L. & Graciela, V. 1988 : Influence of temperature on growth rate & lag phase of fungi isolated from Argentine corn. *Int. J. Food. Microbial* 6 (2) : 179-183. c. f. *Rev. Pl. Path.* (85) 1988 : 110287.

Gowlett J. A.J. 1984 : *Ascent to civilization - The Archaeology of Early man.* Roxby Archaeology Limited, London : pp. 208.

Greathouse, G. A. & Ames, L.M. 1945 : Fabric deterioration by thirteen described and three new species of *Chaetomium. Mycologia* 37 : 138-155.

Gumargalieve, K.Z., Kalinia, I.G., Mironova, S.N. & Moiseevge, Yu. V. 1991 : The effect of nitrates on growth and development of *Phytophthora infestans (Mont.)* de Bary. "*Mikologiya Fitopathologia* 25 (1) 53-56, c. f. *Rev. of Pl. Path,* (1992)" 71 : 675.

Haeckel, E.H. 1866 : *Generelle Morphologie,* 2 vols., Berlin.

Hasija, S.K. 1969 : Sulphur requirements of *Curvularia pallescens* Boed., *Alternaria citri* Cil & Pierce and *A. tenius* Auct. *Mycopath Mycol. App.* 39 (2) : 139-143.

Hawker, L.E. 1950 : *Physiology of Fungi.* Univ. of London, pp. 360.

Hockenhull, D.J.D. 1948 : Studies in Penicillin Production by *Penicillium notatum* in surface culture. 2. Further studies in the metabolism of sulphur. *Biochem. J.* (London) 43 : 498-504.

Hooper-Greenhil, E. 1991 : *Museum and gallery eduction.* Leicester Uni. Press, Leicester and London, pp, 18.

Hueck, H.J. 1968 : The biodeterioration of material – an appreasal., In Bideterioration of materials Elsevies Publ. Co. Ltd., London, pp. 6-12.

Hunter, D. 1957 : *Paper Making, The History and Technique of an Ancient craft.* The Crosset Press, London, pp. 611.

Hyde, H.A. & Williams, D.A. 1961 : Atmospheric pollen and spores as causes of allergic diseases, hay fever, asthma and aerospora. *Advancement of Sciences,* London, 17 : 525.

ICOM 1990 : ICOM statutes code of Professional Ethics, Paris. P. 3 and 23-24.

Janposri, K. 1991 : Fungal growth on paper paintings : Its treatment, removal & prevention. In, *"Biodet. of Cul. Prop."* (Eds. Agrawal, O.P. & Dhawan, S.) MacMillan India Ltd., New Delhi : 28-34.

Jayanti, S., Rai, B. & Singh, V.M. 1992 : Fungal Deterioration of old Books and Journals. *Proc. of 79th ISC,* Botany, Baroda. 11.

Jedrzejewska, H. 1968 : The Damaging Influence of Disinfecting Agents on Sensitive Ancient Materials. In, "Conference on Museum Climatology," (Ed. Thompson, G.) London : International Institute for the conservation of Historic and Artistic works, c.f. *Controlling Museum Fungal Problems* by Strang, T.J.K. & Dawson, J.E. - 1991, C.C.I. Publication, Canada.

Jeyaraj, V. 1991 : Some observation of fungal attack on Museum and Allied Objects. In, *"Biodet. of Cul. Prop."* (Eds. Agrawal, O.P. & Dhawan, S.) MacMillan India Ltd., New Delhi : 104-107.

Joshi, B.R. 1989 : Preservation of Palm-Leaf Manuscripts. *Con. of Cul. Prop.* New Delhi, 22 : 120-127.

Joshi, K.B. 1947 : *Paper Making (As a cottage Industry),* Maganvadi, Wardha, pp. 226.

Keyworth, W.J. 1959 : A modified La Rue cutter for selecting single spores and hyphal tips. *Trans. Brit. Mycol. Soc.* 42 : 53-54.

Khanna, K.K. & Chandra, S. 1972 : Antifungal activities in some Plant extracts. *Proc. Nat. Acad. Sci. India* 42 (B) : 300-302.

Kingsland, G.C. 1985 : Relationship between temperature and survival of *Aspergillus flavus* exfries in naturally contaminated maize grain. *J. stored products Res.* 22 (1) : 29-32.

Kowalik, R. 1980 : Microbiodeterioration of Library Materials, Part-II. Microbiodeterioration of basic organic materials, Ch. IV. *Restaurator,* 4 : 135-219.

Kuć, J. 1995 : Phytoalexins : Stress Metabolism and Resistance in plants. *Ann. Rev. Phytopathol.* 33 : 275-297.

Lal, B. 1977 : The influence of Phosphorus sources on the growth and sporulation of three isolates of *Colletotrichum capscii* (Syd.) Butler & Bisby. *Proc. of Nat. Acad. of Sciences* 47 (3) 169-171.

Lal, B. & Rai, R.N. 1980 : Soft rot of chiku (*Achras sapota* Linn.) caused by a new species of *Phomopsis. Nat. Acad. Sci. Letters,* 3 : 139-140.

Lal, B.B. 1966 : Indian paintings - their techniques, composition & preservation. Proc. of the I Seminar - *Con. of Cul. Prop.,* New Delhi : 26-36.

Lal, B.B. 1972 : Indian Murals : Techniques and conservation. In, "*Con. in Tropics.*" Proc. of Asia-pacific Seminar on *Cons. of Cul. Prop.* (Ed. Agrawal, O.P.) The International centre for the Preservation and the Restoration of Cul. Prop. Rome : 2-19.

Lakshmikant, Kamal & Mathur, S.N. 1990 : Effect of pH on cellulose degrading activity of five cellulolytic fungi. *Proc. of Nat. Acad. of Sci.* 60 B : 357-358.

Lakshmikant & Mathur, S.N. 1991 : Biodeterioration of cultural property by cellulolytic Fungi in, "*Biodet. of Cul. Prop.*" (Eds. Agrawal, O.P. & Dhawan, S.) MacMillan India Ltd., New Delhi : 168-172.

Levi M.P. & Preston, R.D. 1965 : A chemical and microscopic examination of the action of the soft-rot fungus, *Chaetomium globosum* on beech wood. *Holzforschung*, 19 : 183-190.

Lewis, G.D. 1986 : Collections, collectors and Museums—A brief world survey. In, *Manual of Curatorship* (Ed. Thompson J.M.A.), Butteworths, London : 7-22.

Lilly, V.G. & Barnett, H.L. 1951 : *Physiology of the fungi*, Toronto London, Mcgraw, Hill Book & Company INC, New York, pp. 464.

Lodha, B.C. 1964 : Studies on Coprophilous Fungi-I *Chaetomium. J. Ind. Bot. Soc.* 43 : 121-140.

Lokesha, S., Vasantha, K. & Shetty, H.S. 1986 : Effect of plant extracts on growth and sporulation of *Aspergillus flavus. Plant Dis. Res.*, 1 (1-2) : 7981. c.f. : *Rev. Pl. Path.* 67 1988 : 4864.

Laque, J. 1989 : Effect of temperature, Culture medium and pH on colony growth of *Diplodia mutile* Fr. and Mont. *Analg. Del. Jardin Botanico de Madrid* 1989 46 : 215-221. c.f. *Review of Pl. Path* 69 : 7851.

Lyr, H. 1962 : Detoxification of heartwood toxins and chlorophenols by higher fungi. *Nature* (London), 195 : 289-290. c.f. Deterioration of wooden materials by microorganisms by Garg, K.L. and Mukerji, K.G. In, *Recent Advances in Biodeterioration and Biodegradation* (Eds. Garg, K.L., Garg, N. & Mukerji, K.G.), Naya Prokash, Calcutta : 429-480.

Malca, I., Erwin, D.C., Moje, W. & Jones, B. 1966 : Effect of pH and carbon and nitrogen sources on the growth *Verticillium alboatrum. Phytopathol. 56.* : 401-406.

Mandels, G.R. 1954 : Metabolism of sucrose and related oligosaccharides by spores of fungus *Myrothecium verrucaria. Plant Physiol.* 29 : 18-26.

Martinelle, S.D. 1976 : Condition of *Aspergillus nidulans* in submerged culture. *Trans. Brit. Mycol. Soc.* 67 : 121-125.

Mathew, D.S. 1993 : Cultural and Pathological studies of certain soil borne Fungi. Ph. D. Thesis, M.S. University, Baroda. pp. 203.

Mills, J. S. & White, R. 1987 : *The Organic Chemistry of Museum objects.* Butturworths, London. pp. 165.

Mishra, A. & Dhawan, S. 1987 : Fungicides and their Application in Conservation of Cultural Property. *J. Ind. Museums,* 43, New Delhi : 155-162.

Mishra, S.B. & Dixit, S.N. 1976 : Fungicidal spectrum of the leaf extract of *Allium sativum. Indian Phytopath.* 29 : 448.

Mishra, S.B. & Dixit, S.N. 1977 : Fungicidal properties of *Clematis gouriana. Indian Phytopath.* 30 : 577-579.

Mishra, S.B. & Dixit, S.N. 1979 : A simple method for determining fungitoxicity of vapours of plant extracts. *Ind. J. Mycol. Pl. Patho.* 9 (2) : 250-251.

Mishra, G. & Mishra, K.B. 1992 : Aeromycoflora of a mica mine. In, "*National Sym. On. Management of microbes in service of mankind.*" Allahabad Uni., 19-21.

Mislivee, P.B. & Tuite 1970 : Temperature and relative humidity requirements of species of *Penicillium* isolated from yellow dent corn kernels. *Mycologia,* 62 : 74-88.

Montani, M., Graciela V., Silvia, L.R., & Pilar, B. 1989 : Temperature influence on *P. citrinum* growth & citrin accumulation kinetics. *Int. J. Food Microbiol* 7 (2) : 115-122. c.f. *Biological Abstracts* 87 (2) : 22259.

Nair, S.M. 1972 : Biodeterioration of museum materials in tropical countries, In, "*Conservation in tropics.*" Proc. of Asia-Pacific seminars on Con. of Cul. Prop. (Ed. Agrawal, O.P.). The International Centre for the Preservation and the restoration of Cul. Prop., Rome. : 150-158.

Nair, S.M. 1974 : Evaluation of some common Fungicides used for the protection of museum materials. *Con. of Cul. Prop. in India.* 7 : 38-42.

Nair, S.M. 1977 Biodeterioration of Paper. *Con. of Cul. Prop. in India.* New Delhi, 10 : 22-28.

Nair, S.M. 1991 : Problems on Biodeterioration in Natural History Museums. In, "*Biodet. of Cul. Prop.*" (Eds. Agrawal, O.P. & Dhawan, S.) MacMillan India Ltd. New Delhi : 35-44.

Nair, S.M. 1996 : Biodeterioration : Gaps in our knowledge and need for further research. In, "*Recent trends in conservationm of art heritage.* (O.P. Agrawal's Felicitation volume Ed. Dhawan, S.) Agam Kala Prakashan, New Delhi, 181-188.

Narayanan, S.A. 1963-a : Some interesting microfungi from flood affected substrata in Poona, India. *Sydowia,* 16 : 205-211.

Nene, Y.L. & Thapliyal, P.N. 1979 : *Fungicide in Plant disease control.* Oxford and IHB Pub. Co., New Delhi. pp. 507.

Nigam, R.K. & Pathak, N.C. 1991 : Environmental Mycology of a Tannery and a Texitle unit at Kanpur : Allergical Aspect In, "*Biodet. of Cul. Prop.*" (Eds. Agrawal, O.P. & Dhawan, S.) MacMillan India Ltd. New Delhi : 375-386.

Nyuksha Ju P. 1956 : The fungi enhancing book and paper deterioration. Microflora of Books and Paper. *Bot. zh.* 41 (6) : 797-809.

Nyuksha Ju. P. 1994 : The Biodeterioration of paper and Books. In, "*Recent Advancees in Biodeterioration and Biodegradation Vol. I.*" (Eds. Garg, K.L., Garg, N. and Mukerji, K.G.) Naya Prokash, Calcutta : 1-88.

Omelianskii, W.L. 1953 : *Izbrannyie Trudy,* Volume. II, Izdatietstive Akademii S.S.S.R. Moskwa, 380.

Pandey, B.P. 1969-70 : *Taxonomy of Angiosperms.* Jai Prakash Nath & Company, Meerut. pp. 605.

Pandey, R.S., Bhargava, S.N., Shukla, D.N. & Dwivedi, D.K. 1983 : Control of Pestalotia fruit rot of guava by leaf extracts of two medicinal plants. *Revists Mexicana De Fitopatologia,* 2 : 15-16.

Park, D. 1976 : Nitrogen level and cellulose decomposition by fungi. *Int. Biodet. Bull.* 12 : 96-103.

Pericin, D., Jarkak, M., Antov, M., Vujicic, B. & Kevrsesan, S. 1992 : Effect of inorganic phosphate on the secretion of Pectinolytic enzymes by *A. niger., Letters in Applied Microbiology,* 14 (6) : 275-278. c.f. *Rev. of Pl. Pathol.,* 71 : 5251.

Plenderleith, H.J. 1957 : *The Conservation of Antiquities & work of Art. Treatment, Repair & Restoration,* New York, Oxford University Press, London, pp. 394.

Plenderleith, H.J. 1979 : *The Conservation of Antiquities & work of Art. Treatment, Repair & Restoration,* New York, Oxford University Press, London, pp. 394.

Prasad, R. & Sarkar, S. 1977 : Evolutionary extremophilic Archaeal domain of life *Curr. Sci.* 73 : 842-854.

Rai, R.N. 1982 : Pathological & Physiological Studies of certain fungi causing fruit rot disease. D. Phil. Thesis, Allahabad University, pp. 217.

Rajak, R.C. 1983 : Nutritional requirements of two species of *Colletotrichum. Acta. Bot. Indica* 11 (2) : 155-160.

Ranjan, S., Govindjee & Laloraya, M. 1955 : Chromatographic studies on the amino acid metabolism on healthy and diseased leaves of *Croton sparsiflorus. Proc. Nat. Inst. Sci. India.* 22 : 42-47.

Rao, A.S. 1965 : Fungal flora from two soils at Tirupati. *Curr. Sci.,* 34 : 490.

Rebrikova, N.L. & Manturvoskaya 1993 : Study of Factors Facilitating the Loss of Viability of Microscopic Fungi in Library and Museum collections. ICOM Committee for Conservation, 10th Triennial Meeting Washington, D.C. U.S.A. 22-27 August, 1993 : 887-890. c.f. An overview of Problems of Biodeterioration of Cultural Property in Asia by Agrawal, O.P. *Con. of Cul. Property in India,* 1994, 24 : 90-107.

Read, R. 1972 : *Ancient skins, parchments and leather.* Seminar Press, London and New York.

Rebbins, W.J. 1937 : The assimilation by plants of various forms of nitrogen. *Amer. J. Bot.* 24 : 243-250.

Ross, D.J. 1960 : Physiological studies of some common fungi from grassland soils. *N.Z.J. Sci.* 3 : 219-257.

Saccardo, P.A. 1882-1931 : Sylloge Fungorum Omnium Hucusue Cognitorum 25 Vols. Published by the author, Pavia.

Saccardo, P.A. 1886 : Sylloge Fungorum Omnium Hucusue Cognitorum 4 : 1-807, Pavia.

Saini, S.S. & Sidhu, H.K. 1983 : Studies on the carbon nutrition of two spp. of *Choanephora. Geobios.* 10 (4) 149-151.

Saiz-Jimenez, C. 1990 : Analytical approaches for the study of organic compounds in weathered building stones, In : proc. of the Ad. Workshop 'Analytical methodologies for the investigation of damaged stone (ed. F. Veniale U. Zezza) Pavia.

Sarbhoy, A.K. 1965 : Some nutritional studies of the genus *Zygorhynchus Can. J. Microbiol* 11 : 297-308.

Sarkar, H. 1981 : *Museums and Protection of Monuments and Antiquties in India.* Sandeep Prakashan, Delhi : 227.

Sarwal, B.M., Sethi, J.K. & Singh, R. 1982 : A comparative physiological account of two wood-rot fungi I, Carbon nutrition, *Biovigyanam* 8, 2 : 141-143.

Savory, J.G. & Pinion, L.C. 1958 : Chemical aspect of decay of beechwood by *Chaetomium globosum. Holzforschung* 12 (4) : 99.

Seifert, K.A. 1983 : Decay of wood by the Decrymycetales. *Mycologia* 75 (6) : 1011-1018.

Sekhawat, P.S. & Prasad, R. 1971 : Antifungal properties of some plant extracts. (I) Inhibiton of spore germination. *Indian Phytopath.* 24 : 800-802.

Selvaraj, J.C. 1971 : Morphological and physiological studies of isolates of *Verticillium dahliae. Indian Phytopath.* 24 (3) : 471-480.

Sethuramalingam, S., Mohamed, S.S., Palaniappan, R. 1991 : Studies on the Fungal populations from Museum specimens isolation & Identification. In, "*Biodet. of Cul. Prop.* (Eds. Agrawal, O. P. & Dhawan, S.) MacMillan India Ltd., New Delhi", 148-152.

Shah, A.R. 1990 : Application of Cashewnut oil as wood Preservative. *Con. of Cul. Prop. in India, New Delhi* 23 : 28-29.

Shah, N.R. 1990 : Conservation of a wooden plaque. *Con. of Cul. Prop. New Delhi,* 28 : 26-27.

Shah, N.R. & Arya, A. 1995 : Problems of Biodeterioration-A case study. In, "*Conservation, Preservation & Restoration-Traditions, Trends and Techniques.*" (Eds. Kamalakar, G. and Pandit, R.) B.A.C.R. I., Hyderabad : 155-158.

Shah, N.R., Arya, A. & Shah, R.P. 1991 : Growth of fungi on Pavri (a musical instrument) : Study & Treatment. *Con. of Cul. Prop. in India, New Delhi, Silver Jubilee Volume* : 101-102.

Shah, N.R., Shah, R.P. & Arya, A. 1992-93 : Biodeterioration of Flute, a musical instrument-by *Cochliobolus* and *Penicillium. Studies in Museology, M.S. University, Baroda,* 26 : 35-36.

Shah, P.B. 1961 : *Vishnu Dharmottara Purana* (G.O. Series -137) Baroda, Khand-III, 89 : 194.

Shah, R.P., Shah, N.R., Shah, A.R. & Arya, A. 1992 : Biodeterioration of Paintings & Wooden plaque by *Aspergillus & Cladosporium. Geobios New Reports* 11 : 171-172.

Shah, S., Raval, M. & Chhatpar, H.S. 1991 : Studies on Biodeterioration of papers. In, "*Biodet. of Cul. Prop.*" (Eds. Agrawal, O.P. & Dhawan, S.) MacMillan India Ltd., New Delhi., 127-138.

Sharma, B.R.N. 1992 : Pest and Pesticides in Archaeology and Archives. *Con. of Cul. Prop. in India,* 25, New Delhi, 52-55.

Sindall, R.W. 1919 : *The Manufacture of Paper,* Constable & Co. Ltd., 10 & 12 Orango street, Leicester Square W.C. : 276.

Singh, A.P. 1977 : Studies on some soil fungi. D. Phil. Thesis, University of Allahabad. pp. 160.

Singh, B.P. & Chand, J.N. 1973 : Seasonal distribution of rhizosphere fungi in betle gardens. *Sci. & Cult.* 39, c.f. *Fungi of India Part-I* by Bilgrami *et al.* (1979).

Singh, D.B. 1980 : Effect of culture media, pH & temperature on growth behaviour of *Alternaria brassicae and Drechslera graminea, Proc. of the Ind. Nat. Sci. Acad.* 45 B (3) : 393-396.

Singh, R.S. & Tandon, O.P. 1991 : A mycological analysis of the air in the Bharat Kala Bhavan. In, "*Bio. Det. of Cul. Prop.*" (Eds. Agrawal, O.P. & Dhawan, S.) MacMillan India Ltd., New Delhi, 405-411.

Sparrow, F.K. 1958 (1959) : Interrelationships and phylogeny of the aquatic phycomycetes, *Mycologia* 50 : 797-813.

Spieksma, F.T.M. 1980 : Importance of aerobiological studies for the prediction of human diseases, *Aerobiological studies for the prediction of human diseases,* Aerobiol. Symp. Munich. : 307-315.

Srivastava, A.K. 1991 : On the aeromycology of *Chaetomium* in Lucknow. In, "*Biodet. of Cul. Prop.*" (Eds. Agrawal, O.P. & Dhawan, S.) MacMillan India Ltd., New Delhi, 397-404.

Srivastava, A.K. & Singh, K.V. 1982 : (Letter to the Editor) *Indian Drugs,* 19 (6) : 245.

Srivastava, R.C., Tewari, D.K., Katiyar, N. & Lal, B. 1988 : Efficacy of some fungicides against *Chaetospharopsis truncata* var. *indica* Rai & Lal causing banana rot. *Geobios New Reports* 7 : 94-96.

Steinberg, R.A. 1941 : Sulphur and trace element nutrition of *Aspergillus niger. J. Agr. Research,* 63 : 109-127.

Strang, T.J.K. & Dawson, J.E.	1991 :	*Controlling Museum Fungal Problems. Technical Bulletin,* Canadian, Conservation Institute, Canada. 12 : 1-8.
Strzelczyk, A.B. & Leznicka, S.L.	1981 :	The role of fungi & bacteria in consolidation of books., *Int. Biodet. Bull.* 17 (2) : 57-67.
Subbaraman, S.	1993 :	Conservation of Mural Paintings. *Curr. Sci.*, 64 : 736-753.
Sutaria, R.N.	1958 :	*A text book of Systematic Botany.* Khadayata Book Depot, Ahmedabad-1. pp. 693.
Szczepanowska, H.	1986 :	Biodeterioration of art objects on paper. *The Paper conservator.* 10 : 31-39.
Tandon, M.P.	1950 :	Sulphur requirement of *Pestolotia malorum* and *Pestalotia psidii. Proc. Ind. Acad. Sci.* 32 (1) (B) : 7-11.
Tandon, M.P., Bhargava, V. & Jamaluddin	1975 :	Some new fruit rot diseases. *Ind. Phytopath.* 28 : 571-572.
Tandon, R.N.	1961 :	Physiological Studies on Some Pathogenic Fungi, Scientific Research Committee, U.P. Allhabad, pp. 80.
Tandon, R.N.	1967 :	Final Technical Report. Pl-480 Scheme (FG-IN-133). University of Allahabad. pp. 158.
Tandon, R.N. & Bilgrami, K.S	.1958 :	Sulphur requirements of two species of *Phyllosticta Proc. Nat. Acad. Sci. India.* 28 : 293-297.
Thakre, R.P. & Bhajbhuje	1991 :	Biodeterioration of Books & Journals. In, "*Biodet. of Cul. Prop.*" (Eds. Agrawal, O.P. and Dhawan, S.) MacMillan India Ltd., New Delhi : 139-147.
Thind, K.S. & Rawla, G.S.	1958 :	Studies on the nutrition of fungi III : The influence of different sources of carbon on the growth of three Anthracnose fungi. *Proc. Nat. Acad. Sci. India* 28 (B) : 373-378.
Tilak, S.T.	1991 :	Biodeterioration of paintings at Ajanta. In, "*Biodet. of Cul. Prop.*" (Eds. Agrawal, O.P. & Dhawan, S.) MacMillan India Ltd., New Delhi : 307-312.
Tilak, S.T., Sharma, B.R.N., Sengupta, S.R. & Kulkarni, R.L.	1970 :	The deterioration and microbiological studies of Ajanta & Ellora paintings. *Con. of Cul. Prop. in India, Proc. of V. Seminar-* New Delhi : 77-82.

Umalkar, G.V. 1980 : Influence of nitrogen sources on the growth and sporulation in two spp. of *Pestalotia. Botonigue* : 9 (4) : 83-87. c.f. *Rev. of Pl. Pathol.*, 62 (1983) : 2856.

Uniyal, C.P. 1991 : Biodeterioration of Indian miniature paintings - problems and prospectus. In, "*Biodet. of Cul. prop.*" (Eds. Agrawal O.P. and Dhawan S.), MacMillan India Ltd., New Delhi : 91-97.

Upadhyay, M.L. & Gupta, R.C. 1990 : Effect of extracts of some medicinal plants on the growth of *Curvularia lunata. Ind. J. Mycol. Pl. Pathol.* 20 (2) : 144-145.

Urs, K.C.D. & . Reddy, G.V.P 1991 : Mass trapping of American Cockroach *Periplaneta americana L.* (Blattidae : Dictyoptera) in Synthetic Pheromone Trap. In, "*Biodet. of Cul. Prop.*" (Eds. Agrawal, O.P. & Dhawan, S.) MacMillan India Ltd., New Delhi : 478-483.

Venugopal, B. & Menon, U. 1995 : Museums and Community : Some Musings. *Studies in Museology*, M.S. University, Baroda. 28 : 37-44.

Verma, S.C. & Johri, B.N. 1992 : Cellulolysis by different fungi. *Proc. of 79th I.S.C.*, Baroda, Section - Botany : 18.

Whittaker, R.H. 1969 : New concepts of kingdoms of organisms, *Science*, 163 : 150-160.

Williamson, D. 1964 : Pathological and physiological studies of some fungi of stored vegetables and fruits. D. Phil. Thesis, University of Allahabad.

Woese, C.R., Kandler, O. & Wheelis, M.L. 1990 : *Proc. Natl. Acad. Sci.*, USA, 87 : 4576-4579.

Yamamoto, L.A. & Segal, I.H. 1966 : The inorganic sulphate transport system of *Penicillium chrysogenum. Arch. Biochem. Biophys.* 114 : 523-538.

Yu, Clare, C.C. 1954 : The culture and spore germination of *Ascobolus* with emphasis on *Ascobolus magnificus. Am. J. Bot.* 41 : 21-30.

Appendix–1

BIODETERIORATION OF CHINESE SCREEN BY TWO SPECIES OF ASPERGILLUS

– *N.R. Shah, A. R. Shah*[1], *R.P. Shah and Arun Arya*[2]
Department of Museology
Faculty of Fine Arts, M.S. University of Baroda
Vadodara-390002.
[1]Baroda Museum and Picture Gallery, Vadodara 390002.
[2]Botany Department, Faculty of Science,
The M. S. University of Baroda, Vadodara.

Key words : *Biodeterioration, Chinese screen, Aspergillus*

In recent survey of fungal flora of Museum and Picture gallery,

Baroda the authors obtained two new biodeteriogens associated with a wooden chinese screen. Repeated isolations were made during June-July 1990. Pure cultures were obtained in PDA slants. On the basis of morphological characters the fungi have been identified as *Aspergillus ustus* (Bainier) Thom and Church and *A. nidulance* (Eidam) Winter.

Appendix – 2

BIODETERIORATION OF FLUTE, A MUSICAL INSTRUMENT, BY *COCHLIOBOLUS & PENICILLIUM**

– *N.R. Shah, R.P. Shah & Arun Arya*[1]

The flute, a musical instrument has several names like bansi, bansuri, venu, vamishi, kuzhal, murli etc. It is one of the earliest instruments of the Sushira (wind) variety. The origin of the flute is traced to the vedic period. The flute is so much associated with Lord Krishna and, therefore, it is very popular and celebrated instrument all over India.

In ancient India, the flute was very common in the religious music of the Buddhists. This is represented in Indian sculptures at Sanchi. In the sculptures of Amravati and in the paintings at Ajanta and Ellora, the flute is shown as played by human and celestial beings.

Mostly the flute is made of bamboo, but some of the flutes are also made of ivory, sandal wood, ebony, rakta-chandan, iron, bell-metal, silver and gold. But the cheap bamboo flute has been preferred to all these on account of the excellent quality of its tone and its fine resonating properties. The flute consists of a simple cylindrical tube of uniform bore closed at one end. The mouth-hole called the mukha randhra is pierced at a distance of about three fourth of an inch from the closed end (i.e. at the top of the flute). There are six to nine finger holes which are uniform in size. The mouth hole and the finger-holes are arranged in a straight line. The mouth hole is slightly bigger than the finger holes. Different types of flutes are found in modern days, but the shapes remains constant in all these flutes; only their lengths and the number of holes vary. The length can be from eight inches to two and half feet.

This flute is quite different than the other flutes. In this flute mouth hole is in the centre and finger holes are at both the ends. Mouth hole and finger holes are not arranged in a straight line, Mouth hole is at the reverse side of the finger hole, but two sound producing holes and finger holes are in a straight line. This flute is known as "Mohari Pava" by trible community. The size of

---

* Studies in Museology, 26 : 35-37, 1992-1993.

1. Dr. Arun Arya, Lecturer, Dept. of Botany, Faculty of Botany, M.S. University, Baroda.

this flute is also different than the others. The flute under discussion is of 39" in length.

The department of Museology, M.S. University of Baroda has a good collection of musical instruments. As the department of Museology is situated near and bank of river Vishwamitri, during monsoon season the relative humidity goes upto 90%.

In the recent survey of fungal flora of a museum of the Department of Museology, authors found two new biodeteriogens associated with flute, a musical instrument.

Repeated isolations were made during June 91 to February 92 from flute. The object was cleaned with 90% ethanol and when again fungal growth appeared it was isolated. Pure cultures were maintained on Potato Dextrose Agar (PDA) slants. Morphological details of *Penicillium* were recorded on Czapek's agar (Gilman 1959) medium after eight days of incubation.

On the basis of morphological characters and microscopic studies two fungi were identified from flute. They were *Cochliobolus verruculosus* (Tsuda & Ueyama) Sivan (IMI No. 352915) and *Penicillium citrinum.* Thom, (IMI No, 352916).

Treatment :

First the object was cleaned by using soft brush in order to remove dust, dirt and superficial fungal spores. Then the object was cleaned with methylated spirit for the removal of fungus. For complete removal of fungus, it was desirable to fumigate the object as fungal spores were found alive. A fumigation was done by using thymol as fumigant at 50-60°C temperature (150 g/m^3). The treatment was found satisfactory and no further growth of fungus is reported till now.

Acknowledgements :

Authors are thankful to Prof. R.P. Bhatt, Head, Dept. of Botany, M.S. University of Baroda for providing Laboratory Facilities.

One of us N.R.S. is thankful to U.G.C. authority for financial assistance.

REFERENCES

1. Bhowmik, S. & Jani, Mudrika—"Heritage of Musical Instruments" Department of Museums, Gujarat State, Baroda. 1990.
2. Garg, K.L. and Dhawan Shashi—"Microbial deterioration of wooden cultural Property in India. Vol. XVIII to XX, 1985 to 87.
3. Nair, S.M. — *In Cultural contours of India,* Abhinav publication, Jaipur - 1981.
4. Shah, N.R., Arya, A. & Shah, R. P.—Occurence of certain new fungi on museum materials.
5. Shah R.P, Shah, N.R., Shah, A.R. and Arya, A.—"Biodeterioration of paintings and wooden plaque by *Aspergillus and Cladosporium*". *Geobios New Reports,* Vol. 11, No 2. 1992.

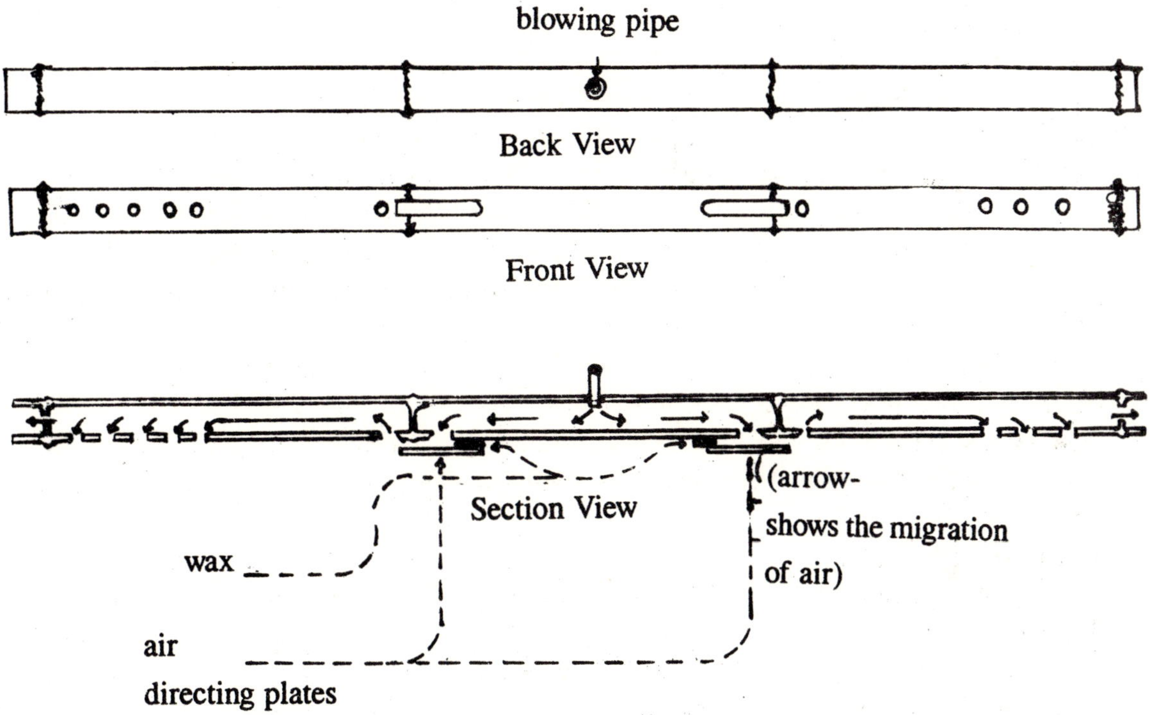

Flute (Mohari Pava)

Appendix – 3

PROBLEMS OF BIODETERIORATION – A CASE STUDY*

– *N. R. Shah and Arun Arya*[1]

Abstract

Deterioration caused by fungi and insects on different organic objects of the department of Museology, M.S. University, Baroda (Gujarat, India) and morphological characters of different fungi are recorded. Various curative and preventive measures of these fungal attacks are suggested.

Introduction

The Department of Museology is a pioneering institution in the field of museum training at the post-graduate level associated with the M.S. University, Baroda (Gujarat, India). The department was established in the year 1952 in the building of Museum and Picture Gallery, therefore, students used to have the advantage of Museum and Picture Gallery. The department is presently housed in a new 3 storeyed building in the campus of Fine Arts opposite Sayaji Baug, near river Vishwamitri. Although the department started in its new building in 1976, the museum was thrown open to public in the year 1989 only.

This museum possesses different types of objects such as potteries, wood carvings, objects of papier mache, metal objects, paintings, paper manuscripts, beadwork, natural history specimens and a remarkable collection of the sculptures of Saptamatrikas. Moreover, the museum has some reserve collection on the first floor[1].

Climatic condition

Department of Museology is situated near the bank of river Vishwamitri. Hence, more dampness is found in it as compared to other places. In monsoon season the relative humidity reaches upto 90 to 95%. Insect activity and growth of microorganisms are observed at frequent intervals. Day to day record of temperature and relative humidity is maintained in order to keep track of the rapid fluctuations in temperature and relative humidity.

Problems of biodeterioration

It is an established fact that the organic objects are more susceptible to the biodeteriorating agencies than inorganic objects. A scientific study was undertaken to find out biodeteriogens and thereafter some curative and preventive measures were taken.

---

* Conservation, Preservation & Restoration, Traditions, Trends and Techniques, 155-158. (Edts. Dr. G. Kmalakar & Dr. V. Pandit Rao) BACRI, Hyderabad, 1995.

1. Dept. of Botany, Faculty of Science. M.S. University, Baroda.

Attack of Fungi

The fungi were isolated from the musical instruments, paper manuscripts and wooden objects by repeated isolations. First the objects were cleaned with 90% ethanol and when fungal growth appeared again, fungi were isolated. Pure cultures were maintained on PDA slants. Final confirmation of fungi was done by IMI, Kew, England. List of fungi isolated from different objects is given in Table 1.

Table-1. Fungi isolated from different objects

Name of the object	*Fungi*	*Material of the object*
1. Miniature painting	i) *Aspergillus tamarii* Kita; IMI No. 334062 ii) *Cladosporium herbarum* (Pres.) Link.; IMI No. 334058	Paper
2. Wooden plaque	*Aspergillus niger* Van Tieghem; IMI No. 334057	Wood
3. Pavri A musical instrument	i) *Cochliobolus pallescens* (Tsuda & Ueyama) Sivan. ii) *Phoma sp.* iii) *Fusarium semitectum* Berk. & Rav.	Wooden part of the musical instrument[2]
4. Flute A musical instrument	i) *Cochliobolus verruculosus* (Tsuda & Ueyama) Sivan. IMI No. 352915 ii) *Penicillium citrinum* Thom.; IMI No. 352916	Wooden part of the musical instrument.[3]
5. Tambura A musical instrument	*Emericella nidulance* (Eidam) Vuill.; IMI No. 352923	Wooden part of the musical instrument.
6. Parsi manuscripts	*Cochliobolus lunatus* Nelson & Haasis; IMI No. 352925	Paper material

Morphological characters of the Fungi

1. *Aspergillus tamarii* : Colonies on Czapek's agar spreading broadly with vegetative hyphae submerged, fruiting areas orange-yellow to brown, in old colonies, light brownish-olive. Reverse uncoloured or occasionally pinkish. Conidia pyriform to globose, 5, 6 or upto 8 µm in diameter.

2. *Cladosporium herbarum* : Colonies greenish black on PDA. Conidiophores erect, branched, septate 5-10 µm in diam. Conidia terminal on short knee-like swellings, single or in chains, elongate, oval, one celled, cylindrical with one to four septa, olive green.

3. *Curvularia (Cochliobolus)* : Mycelium branched, septate, subhyaline or brown, conidiophores unbranched, septate. Conidia acrogenous verticillate or spirally arranged, olivaceous or brown, elliposid or cylindrical, curved or bent 3 or 4 septate one of the central cells being distinctly larger and darker than the terminal cells.

4. *Phoma* : Fruiting body a pycnidium globose or slightly lens shaped with a small papilla at the apex. Spore, small, egg shaped, one celled, hyaline usually with two oil drops, conidiophores thread like or short.

5. *Fusarium semitectum* : Conidia powdery, white to rosy or brown, floccose, aerial mycelium or united into balls in false heads. Macroconidia 3-4 septate occasionally 6 to 10 septate. Larger conidia spindle, straight or slightly curved.

6. *Emericella* : Fruiting body a closed ascocarp-cliestothecium, ellipsoid to pyriform, surrounded by loosely arranged swollen refractive thick walled cells. Peridium membranous of several layers. Asci irregularly arranged, globose to ellipsoid, 8 spores, red brown.

Insect activity

In the show cases of the museum insect activity was observed. Damage to some paper materials as well as to labels was noticed. It was due to the activity of Silver fish (*Lepisma saccharina*) and Book lice (*Liposcelis* sp.). These insects are known for surface damage of the paper materials[4]. Silver fish is also known to cause damage to India ink of the labels of objects on display. Particularly in this museum damage caused to ink of the labels was observed for the first time.

Curative measures

The objects which were affected by fungi were first treated with absolute Alcohol to remove the fluffy fungal growth from the surface of the objects. After that, the objects were fumigated by using 5% solution of thymol in fumigation chamber. The objects which were affected by insects were fumigated by using paradichlorobenzene.

After the above treatments, no further attack of fungi or insect was noticed. The objects were protected from photo-chemical deterioration by using U.V. filters at the source of natural light. The intensity of light and U.V. rays was controlled upto the tolerable limits.

Preventive measures

It is said that 'Prevention is better than cure'. It is recommended to make the show cases air tight to prevent the wide fluctuation of climate. A good house-keeping habit will help in saving the museum materials from biodeterioration. A regular cleaning of the show cases and periodical fungicidal spraying of 0.25% Pentachlorophenol or 0.2% mercuric chloride as well as insecticidal spraying of 5% DDT or Pyrethrum extract is recommended to prevent attack of fungus or insects. Keeping of repellents like paradichlorobenzene, or P.B.C. mixture (equal quantity of paradichlorobenzene, benzene and creosote oil) is extremely essential to keep the insects away.

REFERENCES

1. A. Iftikhar, Conservation of Cultural Property in India. Vol. XVI & XVII, New Delhi, 1983 & 1984.
2. N.R. Shah, et al, Conservation of Cultural Property in India, New Delhi, 1992, pp. 101-102.
3. N.R. Shah, et al, Studies in Museology, 26, Baroda, 1992-93, pp. 25-37.
4. A. Iftikhar, *op. cit.*

Appendix – 4

A STUDY OF ATMOSPHERIC FUNGAL FLORA AND BIODETERIORATION IN DIFFERNT MUSEUMS OF BARODA*

– *N.R. Shah, Arun Arya[1], A.R. Shah[2] and S.K. Bhowmik[2]*
Department of Museology. Faculty of Fine Arts
The M.S. University of Baroda. Vadodara-390002 (Gujarat).

Museums are not merely a store house but they are centres for education in the service of society. The Baroda Museum cum Picture Gallery is famous for its collection of European oil paintings all over Asia. The buildings of museum are situated in pastrol environment of Sayaji Park. This museum is built in the Indo Saracenic style of architecture. Department of Museology is situated near the bank of river Vishwamitri. Its museum possesses different types of objects such as wood carvings, objects of papier mache, paintings, paper manuscripts etc. Oriental institute is situated on the ground floor of Smt. Hansa Mehta Library in M.S. University. All these three spots are situated within a radius of 1 km. in Sayaji Gunj area.

Air spora forms a normal component of the earth's atmosphere. Aerobiological studies have been assuming awareness due to their application in various fields including in clinical allergy (Edmonds and Bennighoof 1977). These studies are helpful in assessing the deterioration and management of environment. The aerobiological data may be used for the formulation of disease forecasting system of plants (Mishra 1992) and human beings (Davies and Smith 1973 and Spieksma 1980). Role of some of these fungi in biodeterioration of museum objects in different museums is emphasised.

Materials and Methods

Various methods have been suggested to isolate the aeromycoflora. In our study, use of culture plate exposure method has been made as gravity slide exposure method is found to have certain drawbacks (Nigam and Pathak 1991) Like :-

---

* Perspectives in Biological Sciences (Edts. V. Rai, M.L. Naik, C. Manoharacharrya) School of Life Science, P.T. Ravishankar Shukla University, Raipur, 1996, pp. 189-196.

1. Department of Botony, M.S. University of Baroda-390002.

2. Museum and Picture Gallery, Baroda.

i) Similar spores are not identified upto generic level (e.g. *Aspergillus, Penicillium* and *Mucor* etc.)

ii) Some spores are too small or too transparent or do not have sufficient individual characteristics e.g. yeast an *Phoma* etc.

iii) Fungal hyphae are not identified.

iv) Many spores can not be trapped due to strong wind.

A mycological analysis of air in the Baroda Museum was done at 6 places i.e. Indian art, Nepalese and Tibetan Galleries, Islamic, Picture and Ethnological galleries and laboratory. In Deptt. of Museology sampling was done on ground floor (Museum). 1st and 2nd floor and in Oriental Institute from main hall.

Five petriplates (100 mm) each were exposed at 1 meter height at all above places for 10 min. They were incubated at 25± 1^{0}C for 7 days. As soon as the colonies were developed they were isolated and pure cultures were maintained onPDA slants. *Aspergillus* and *Penicillium* spp. were identified by growing in Czapek's Dox medium and studying their morphological characters (Bessey 1961, Barnett 1967). Fungi from wood were isolated as per the method suggested by Banerjee and Levy (1970).

Result and Discussion : Aeromycoflora of Baroda Museum

In our studies spores of 22 species belonging to 12 genera were isolated. Of these 1 belonged to Zygomycotina, 11 to Ascomycotina and 10 to Deuteromycotina group. Of the 6 places where petriplates were kept, picture and Indian art gelleries harboured maximum (175 & 162) colonies. Of all these species *Aspergillus niger* and *Penicillium citrinum* constituted about 10% and 8% of the total colonies respectively (Table 1). Isolations from exposed petriplates resulted in pure culture of 5 *Aspergillus* and 2 *Penicillium* spp. One perfect state of *Aspergillus* i.e. *Emericella nidulans* was also isolated. These are common spp. causing biodeterioration of various museum objects. According to Nair (1971) "Fungi that can grow on any suitable substrata play a prominent role in the deterioration of museum materials." The starch and glue used in museum objects may be utilized as raw material by the fungi. They need only 7-9% of moisture percentage of the museum materials to grow. The minimum and maximum relative humidity in the month of isolation was 43 and 52% respectively and average temperature was 25.6^{0}C.

Table - 1. Environmental mycological data in different Galleries of Baroda Museum in Jan. 1994.

Sl.	Fungi	IA	NG	IG	PG	EG	Lab.	Total	%
1.	Alternaria alternata (Fr.) Keisslar	5	-	-	-	-	5	10	1.22
2.	Aspergillus flavus Link	5	5	1	5	10	10	36	4.42
3.	A niger van Tiegh	10	20	-20	10	20	80	9.82	9.82
4.	A parasiticus Speare	10	10	10	-	-	10	40	4.91
5.	A sydowii (Bain & Sart) Thom & Church	5	5	10	10	5	5	40	4.91
6.	A versicolor (Vuill.) Tiraboschi	-	-	10	20	10	5	45	5.52
7.	Cladosporium herbarum (Pres.) Link.	2	2	5	-	-	-	09	1.10
8.	Chaetomium gangligerum L. Ames	-	-	10	10	-	-	20	2.45
9.	C. globosum Kunze and Schm.	5	-	-	-	-	10	15	1.84
10.	C. indicum Corda	-	-	-	10	-	2	12	1.47
11.	Curvularia lunata (Wakker) Boedijn	15	-	10	5	10	2	42	5.15
12.	Emericella nidulans (Eldam) Vuill.	10	-	5	10	-	5	30	3.68
13.	Fusarium monitiforme J. Sheld.	10	10	-	-	10	-	30	3.68
14.	F. oxysprum Schlecht.	10	-	20	5	10	-	45	5.52
15.	Fusarium sp.	5	5	-	-	-	-	10	1.22
16.	Gliocladium sp.	5	-	10	-	-	-	15	1.84
17.	Monilia sitophila (Mont.) Sacc.	10	5	5	5	-	-	25	3.07
18.	Penicillium citrinum Thom.	20	5	20	20	-	-	65	7.98
19.	P. nigricans (Bain) Thom.	10	-	10	10	10	10	50	6.14
20.	Rhizopus stoloniter (Ehr. ex Fr.) Lind.	10	20	-	20	-	-	60	7.37
21.	Trichoderma lignorum (Tode) Harz.	-	-	-	10	10	10	20	2.45
22.	T. viride Pers, ex. Fr.	-	-	-	-	10	10	20	2.45
23.	Sterile white mycelium	10	10	10	5	5	5	45	5.52
24.	Sterile black mycelium	5	5	10	10	5	5	45	5.52
	Total	162	103	146	175	115	114	814	

Note : The data represent the total colonies produced after 7 days on 5 petriplates.

Aeromycoflora of Museology Department and Oriental Institute

Tables 2 & 3 show that spores of 12 different fungi were present in Museology Department and only 4 fungi were isolated in petriplates exposed in Oriental Institute. In Museology Department maximum number of colonies were obtained in the Museum on ground floor while least on 2nd floor. Occurrence of three rare fungi, *Chaetomium globosum, Monoascus ruber* and *Syncephalastrum racemosum* are a feature of special interest.

Table-2 : Environmental mycological data in different places of Department of Museology in Jan 1994.

S.No.	Fungi	GF	IF	IIF	Total	%
1.	Achaetomium globosum	5	-	2	07	5.5
2.	Aspergillus fumigatus	5	5	-	10	7.8
3.	A. niger	10	15	10	35	27.5
4.	A. ustus	5	2	1	08	6.2
5.	Cochliobolus pallescens	5	5	-	10	7.8
6.	C. verruculosus	-	2	-	02	1.5
7.	Emericella variecolor	5	5	5	15	11.8
8.	Monoascus ruber	2	-	-	02	1.5
9.	Penicillium sp.	5	5	2	12	9.4
10.	P. griseofulvum	2	5	5	12	9.4
11.	Rhizopus sp.	5	2	5	12	9.4
12.	Syncephalastrum racemosum	2	-	-	02	1.5
Total 12 spp		51	46	30	127	

Table-3 : Fungi occuring in atmosphere of Oriental Institute

Sl. No.	Fungi	Total colonies produced in 5 petriplates	%
1.	Aspergillus sp.	5	17.2
2.	A. niger	12	41.3
3.	Cochliobolus verruculosus	2	6.8
4.	Emericella veriecolor	10	34.4
	Total 4 Spp.	29	99.7

In Oriental Institute 29 colonies of 4 different fungi i.e. *Aspergillus sp. A niger, Cochliobolus verruculosus* and *Emercella veriecolor* were obtained. Various Aspergilli may cause allergy and aspergillosis in man (Table 4). Many of these spores may remain active for 30-40 years, thus causing a serious health hazard.

Some fungi are cellulolytic like *Chaetomium, Cladosporium Fusarium* and

Trichoderma etc. Table 5 points out some of the most comon symptoms of biodeterioration (Nair, 1980). Arya et.al. (1988) reported spoilage of Rajasthani miniature painting due to *Chaetomium gangeligerum*. Another species of this genera *C. globosum* has been found causing deterioration of some photographs placed in laboratory of the Museum.

Table-4 : Potential allergens found in aeromycoflora of Baroda Museum and Department of Museology.

Sl. No.	*Baroda Museum*	*Deptt of Museology*
1.	Aspergillus flavus	Aspergillus fumigatus
2.	A. sydowii	A. niger
3.	A. versicolor	Aspergillus sp.
4.	Cladosporium herbarum	Penicillium sp.
5.	Emericella nidulans	Rhizopus sp.
6.	Monilia sitophila	
7.	Penicillium sp.	
8.	Rhizopus stolonifer	

Table-5 : Symptoms of biodeterioration of different kinds of Museum materials (Nair 1980)

Sl. No.	*Symptoms*	*Examples*
1.	Stains and discoloration	Caused by surface growth of fungi. on paper, leather, textile etc.
2.	Disfigurement	Positive changes in the structure of a painting due to fungus.
3.	Fibrillization	Loss of coherence of fibrous materials as in wool attacked by microorganisms.
4.	Impenetrations	The invasion of the lumen of certain natural fibres by fungi.
5.	Changes in mechanical properties	Loss of weight and strength of wood or reduction of breaking strength of coffon fibres attacked by fungi.
6.	Changes in chemical properties	Break down of cellulose by microorganisms, foxing on paper, etc.
7.	Development of odour	Mildewing on paper.

A systematic identification of fungi growing on paper materials is done by Dhawan and Agrawal (1986). *Chaetomium* spp. cause consolidation of books (leaves of books are glued together due to over growth of fungus) and a study has been undertaken by Srivastava (1991) from Lucknow and Strezelczyk and Leznicka (1981) from Poland. Lakshmikant and Mathu (1991) reported 33 different fungi from manuscripts, books, woodcrafts, paintings and wall hangings. It included spp. like *Chaetomium spinosum*. *Paecilomyces* and *Scopulariopsis* etc. Various spp. were found in mildewed cellulosic articles like cloth, cotton, paper, siezed cloth etc. (Bagool 1992). Jeyaraj (1991) found *Ascotricha charterum* and *Aspergillus niger* on books. In our studies also *A. niger* was obtained from books in Oriental Institute.

Greathouse and Ames (1945) described fabric deterioration by 13 described and 3 new spp. of *Chaetomium*. According to Kowalik (1980), development of fungi on paper depends on pH of the object.

The fungal attack occurs in museum objects which are generally located in moist conditions. The actual conditions in which faster fungal growth appear are as follows :

a) Supply of food materials from which the fungus can derive its nourishment.
b) Suitable temperature.
c) Supply of moisture.
d) Supply of oxygen for the respiration of the fungus.
e) Presence of some original infection.

Plant Pathogenic Fungi

In our investigation many fungi that are recorded are the weak plant pathogens like *Alternaria alternata, Cladosporium herbarum, Fusarium moniliforme, F. oxysporum* and *Rhizopus stolonifee*, their occurrence may be due to vestation and debris available in the surroundings of Sayaji Park which possesses a variety of plant species. The fungal spores may be blown from these sources and spread in museum environment. The occurance of these spores have been reported from various places like Dehradun (Kumar 1984). Kanpur (Nigam and Pathak, 1991), Lucknow (Khandelwal, 1991). Benaras (Singh and Tandon 1991), Kumaon (Chauhan and Sharma 1991) and Jabalpur (Mishra 1992). The spp. of *Alternaria, Cladosporium* and *Curvularia* are reported from mica mine at Gaya, Bihar (Mishra & Mishra 1992). *Trichochecium roseum* was reported from poland (strzelczyk, 1976). It is interesting to note that certain fungi were found to cause deterioration of museum objects,chiefly on miniature paintings and musical instrument (Table 6 and 7) Shah *et al.* (1992) reported occurrence of *Aspergillus niger* on wooden plaque depicting God Murugan.

Fungi, bacteria and actinomycetes are the major contributors to the daterioration of cultural property in tropical conditions. As recorded above the spore diversity is very high. The presence of these fungal inocula in the atmosphere clearly warrants immediate protective measures and reduction of humidity levels in the display chambers to avoid fresh attacks on organic materials. Allergenic and pathogenic nature of some airborne spores may pose potential hazards to visitors of these Museums.

Table-6 : Fungi reported from the objects belonging to Baroda Museum

1.	*Aspergillus terreus*	Rajasthani miniature Painting (paper)	Arya et al.	1988
2.	*A. versicolor*	"	"	"
3.	*Chaetomium gangligerum*	"	"	"
4.	*Aspergillus nidulans*	Chinese screen (wooden)	Shah et al.	1993
5.	*A. ustus*	"	"	"
6.	Chaetomium globousum	On paintings & Photographs	Shah et al.	(in press)

Table 7 : Fungi reported from the objects belonging to Department of Museology

Sr. No.	Organism	Object	Reference
1.	*Cochliobous verruculosus*	Flute	Shah et al. 1992-93
2.	*Penicillium citrinum*	"	" "
3.	*Cochliobous pallescens*	Pavri	Shah et al. 1992
4.	*Phoma sp.*	"	" "
5.	*Fusarium ?semitectum*	"	" "
6.	*Aspergillus tamarii*	Miniature painting	Shah et al. 1992
7.	*Cladosporium herbarum*	"	" "
8.	*A. niger*	Wooden plaque	" "

Acknowledgements

Thanks are due to Dr. B.C. Sutton, Director, IMI, Kew, England for confirming the identity of fungi and to Prof. R.P. Bhatt, Head Department of Botany, Faculty of Science, The M.S. University of Baroda for providing the laboratory facilities.

REFERENCES

Arya A. S.K. Bhowmik and A.R. Shah. 1988. Biodeterioration of paintings. *Geobios new Reports* 7 : 179-180.

Bangool, R.G. 1992 Cellulose-degrading fungi around Bombay. *J. Bombay Nat. Hist. Society* 90 : 62-64.

Banerjee, A.K. and J. F. Levy. 1970 Techniques for the isolation of fungi from wood. *Int. Biodeth Bull.* 6 : 37-41.

Barnett, H.L. 1967 *Illustrated genera of Fungi Imperfecti*, Burgers Publ. Co. Minneapolis.

Bassey, E.A. 1961 *Morphology and Taxonomy of Fungi*, Hafner.

Chauhan. M.S. and Chaya Sharma, 1991. In: Biodeterioration of Cultural Property. eds. O.P. Agrawal and S. Dhawan by MacMillan India Ltd., 412-426.

Davies, R.R. and L.P. Smith. 1973 Aspects of Human allergy and fungal spores. *Clinical Allergy* 3:263-267.

Dhawan, S. and O.P. Agrawal 1986. "Fungal flora of miniature paper paintings and lithographs," Int. Biogs. Bull. 22: 95-99. Edmonds. R.L. and W.S. Bennighoot I.B.P. Aerobiology Programme pp. 17.

Greathouse, G.A. and L.M. Ames. 1945 Fabric deterioration by thirteen described and three new species of Chaetomium. *Mycologia* 37 : 138-155.

Jeyaraj, V. 1991. In : *Biodeterioration of Cultural Property*, Macmillan India Ltd. 104-107.

Khandelwal, Asha. 1991. In : *Biodeterioration of Cultural Property*. Macmillan India Ltd. 387-386.

Kowalik, R. 1980 "Microbiodeterioration of library materials Part II, Microbiodeterioration of basic organic materials", Ch. IV. *Rastaurator* 4 : 135-219.

Lakshmikant and S. Mathur. 1991. In : *Biodeterioration of Cultural Property*. Publ. by Macmillan India Ltd., 168-172.

Mishra, R.P. 1992 Aerobilogical studies of two locations in Jabalpur (M.P.) I. Winter season *Biome* 5 : 30-38.

Mishra, G. and K.B. Mishra 1992. In : *Symp. Management of microbes*, Alld. Univ. 19-21.

Nair, S.M. 1971, In : *Biodeterioration of materials* (eds) A.H. Walters & E.H.H. Plas, Applied Science, London U.K. 2:401-421.

Nair, S.M. 1980. In : *Conservation in tropics* ed. O.P. Agrawal 150 pp.

Nigam, R.K. and N.C. Pathak, 1994 : Environmental Micology of a Tannery and a Textile unit at Kanpur : Allergical Aspects. In : *Biodeterioration of Cultural Property*. Publ. by MacMillan India Ltd. 375-386.

Shah, R.P. Shah and A. Arya 1992. Biodeterioration of Paintings and Wooden plaque by *Aspergillus and Cladosporium. Geobios new reports* **11**:171-172.

Singh, R.S. and O.P. Tandon 1991. In : *Biodeterioration of Cultural Property* by macmillan India Ltd., 405-411.

Spieksma, F.T.M. 1980 Importance of Aerobiological studies for the prediction of human diseases. *Aerobiol. Symp. Munich.* 308-315.

Srivastava, A.K. 1991. In : *Biodeterioration of Cultural Property* by MacMillan India Ltd. 397-404.

Strzelczyk. A.B. 1976. Adaptation of Fungicides of fungi damaging paper. *Int. Biodetn. Bull* **12** (1) : 19-26.

Strzelczyk, A.K. and S. Leznicka 1981. The Role of Fungi and Bacteria in the consolidation of Book. *Int. Biodent. Bull* **17** (2) : 57-67.

☆☆☆

Index

A

Absidia—21, 46
Absidia spinosa—43
Acetamide—106
Acorus calamus—152
Acremonium—24
Actinomycetes—127
Adjustment of pH—33, 34
Adjustment of temperature—35
Aegelinine—154
Aegle marmelos—154
Aerobiological studies—128
Aeromycoflora—35, 127
Aerospora—10
Ajanta caves—9, 10, 127
Algae—144
Allergens—127
Alternaria—6, 10, 24
A. brassicae—106
Amino acids—96
Anacardium occcidentale—151
Antifungal chemicals—146
Antiquarian law—2
Archaea—16
Archaeo bacteria—16
Ascomycetes—18
Ascomycotina—20, 22
Ascotricha charterun—128
Asiatic society—3
Aspergillus—6, 10, 22, 153
A. flavus—37, 46
A. fumigatus—43
A. japonicus—41
A. nidulans—42
A. niger—41, 42, 43, 117, 148, 150
A. oryzae—42
A. ustus—13, 30, 42, 46
A versicolor—42
Asthana & Hawker's medium-A—49
Aureobasidium pullulans—6
Azadirachta indica—137, 151
Azadirachtin—152

B

Basal medium—47
Basidiomycetes—18
Basidiomycotina—20
Bat—16, 154
Bavistin—147, 161
Betula utilis—139
Bharat Kala Bhavan—127
Bhoj-Patra—139
Biochemical Studies—38
Biocide—7
Biodegradation—15
Biodeteriogens—5, 15
Biodeterioration—3, 4, 15, 127, 144
Biodeterioration effects—4
Biodeterioration of monuments—9
Biodeterioration of papers—39, 137, 140
Biodeterioration of woods—39, 134
Biological agents—4
Birds—16
Book mycoflora—5
Botanical pesticides—151
Botrytis—25
British museum—3
Brown rot—7, 9

C

Calixin—147, 161
Carbohydrates—159
Carbohydrate utilization—81
Carbon nutrition—69
Carbon sources—73
Carboxy methyl cellulose—80
Cashewnut oil—151
Cedrus deodara—137
Cellulase—5, 140
Cellulose—5, 8, 69, 139
Coelomycetes—21
Chaetomium sp.—20, 23, 42, 127
C. cymbiforme—13, 30, 42
C. gangligerum—42
C. globosum—7, 144
Chaetosphaeropsis—109
Chaitya caves—134
Chemical control—148
Chinese screen—12
Chromatograghic studies—38
Cryptogams—9
Cineole—153
Citric acid—81
Citrus medica—154

Cladosporium—6, 10, 25
C. herbarum—42
Cochiliobolus—20, 26, 45
C. verruculosus—12, 30, 42, 147, 148
C. pallescens—41
Colletotrichum dematium—106
C. gloeosporioides—106
Colony diametre method—38
Coniophora—20
Coniophora puteana—7
Conservation—144, 159
Control—39
Control studies—13, 144
Coon's medium—49
Copper sulphate—109
Creosote oil—150. 161
Cultural heritage—1, 2, 163
Cultural property—10
Cultural studies—13
Culture collection—37
Culture medium—31, 47
Culture plate method—35, 128
Cunninghamella—20
Curvularia—10, 25
Cyanobacteria—3
Cyperus papyrus—139
Czapek's medium—49

D

Decrymyces—20
Decrymyces stillatus—7
Decrymycetales—137
Deodar—137
Deterioration—2, 3, 10
Deterioration of paper—5
Deterioration of wood—6
Deuteromycetes—18, 19
Deuteromycotina—20, 24
Dextrin—95
D-fructose—93
D.glucose—93
Disaccharide—70
Discolouration—5
Disease forecasting—128
Disfigurement—145
Donkiporia—7, 20
Dox's medium—49
Drechslera—10, 26
Dry rot—6
Dry weight method—38

E

Egyptian papyrus—139
Elliott's medium—49
Ellora caves—9, 10
Enzymes—8
Eucalyptus—153, 162
Eucalyptus oil—152
Eucalyptus occidentalis—152
Eukaryotic—17
Eupenicillium—23
Euthiotrophic—108
Eutrophic—9

F

Ferrous sulphate—160
Flute—148
Formalin—149
Foxing—6, 145
Frescoes—10
Fructose 79
Fumigation 145
Fusarium—10, 26
F. semitectum—41
F. udum—106
Fungal enzymes—136
Fungal spores—134
Fungi—10, 18
Fungi isolated—41-43

G

Garlic—153
Georgian state museum—144
Geotrichum sp.—43, 46
Gibberella—27
Glucose—68
Godavaj powder—152
Gravity slide method—36, 128
Growth assessment—38

H

Herbal pesticides—152
Hajira tomb—16
Harappa—135
Hexose—69
Human health—14
Hydrogen ion concentration-influence—53
Hypha—11
Hyphomyces—27
Hyphomycetes—21

I

ICOM—1
Identification—37

Indian Museum Movement—3
Indoor aeromycoflora—35, 127, 130, 160
Indus valley civilization—135
Insects—16
Inulin—80
Isolation technique—34, 44

J

Japanese tissue paper—141, 142
Jogimara caves—9

K

Kaolin—139
Kutch Museum—46

L

Lactose—79, 94
L-aspiragine—106
Lalit Kala Academi—144
Lantana—154, 162
Lantana camara—154
Leaf extract—153
Leather objects—151
Lemon—154
Library-Smt.H.M.—128
Lichen—8
Lignin—69
Lithobionts—16
Lithograph—128, 141

M

Maltose—79, 94
Magnesium sulphate—109, 160
Mangifera indica—137
Mango—137
Mastigomycotina—20
Mechanical biodeterioration—8
Mentha arvensis—154
Menthol—154
Mercuric chloride—149
Merulius lacrymans—6
Mesophiles—11, 62
Mica—16
Micro-organisms—127
Miniature paintings—12, 46, 128, 141
Modified Asthana & Hawker's medium-A—49
Monera—17
Monilia sitophila—140
Monosaccharide—70
Monument preservation act—2
Mucor—20, 21
Mucorales—140
Mural paintings—9
Museology dept.—45, 128
Museum—1
Museum collection—2
Museum function—2
Museum & Picture gallery—45, 128
Mushrooms—18
Musical instrument—12
Mycelium—11
Myceteae—18
Mycophyta—17
Mycota—19
Mystox-B—150
Myxomycota—19, 20
Myxotrichum deflexum—96

N

Nephthalene—150
Natural media—32
Natural plant products—151
Nectria—20
Neem—137
Neem oil—151
Nepalese tissue paper—142
Nigrospora—27
Nitrate—106
Nitrite—106
Nitrogen assimilation—96
Nitrogen sources-effect—96

O

Oil refinery—15
Oligotrophic—9
Organic acids—70, 80
Oriental Institute—128
Oxalic acid—80

P

Paecilomyces—10
Palm leaves—139
Paper—5, 161
Paper manufacture—139
Paper manuscript—12
Paradichlorobenzen—148
Parathiotrophic—108
Paranitrophenol—151
Penicillium—6, 23, 45
P. chrysogenum—107
P. citrinum—12, 30, 41, 42, 43, 62, 137
P. funiculosum—150
Pentachlorophenol—150, 161
Pestalotia malorum 108
Pestalotia psidii—153
Phellinus—20
P. contiguuas—7

Phoma—27, 45
Phomopsis vitiola—153
Phosphrorus sources—117
Phosphorus sources-effects—117, 160
Phosphorylation—117
Phudina—154
Phycomycetes—18, 19
Pigeon—16, 144
Plant pathogenic fungi—130
Pleurotus ostreatus—7
Polysaccharide—70
Pollution—15
Poria plaenta—7
Potassium dihydrogen phosphate—118
Potassium hydrogen phosphate—118
Potassium nitrate—160
Potato Dextrose Agar—32
Preventive measures—145, 161
Preparation of Medium—33
Prions—17
Progenote—16
Prokarytic—17
Proteases—96
Protista—17
Psyechrophiles—11, 62
Purification—36
Pyramids—15
Pyrenophora—26

R

Rats—16, 144
Remedial measures—161, 146
RH—11
Richard's medium—49, 159
Rhizopus—20, 21

S

Sal—137
Sanchi—2
Sclerotinia—25
Semi synthetic media—32
Serpula lacrymans—6
Shorea robusta—137
Single spore culture—36
Soft rot—7
Soft rot fungi—144
Sordaria—24
Stachybotrys—28
Stains—145
Starch—80, 95
Stemphylium—28
Stone cancer—15
Sucrose—79, 93
Sulphate sulphur—107
Sulphides—107
Sulphur sources—108
Sulphur surces effects—107
Sweet flag—152
Syncephalastrum—20, 21
S. racemosum—42, 150
Synthetic medium—32, 47
Systemic fungicide—148

T

Tajmahal—15
Talaromyces—23
Teak—137
Tectona grandis—137
Tempera paintings—10
temperature-extreme—61
Temperature-influence—61, 67
Termite—4, 16
Thermophiles—11, 61
Thermotolerent—61
Thiourea—106, 109
Thymol—148
Timber—136
Timber decay—136
Trichoderma—10, 29, 43
T. Kuningii—42
T. ressei—127
T. viride—127

U

Ulocladium—29
Urea—106

W

Wet rot—7
Wood—161
Wood deterioration—136
Wood rotting fungi—135
Wood staining fungi—135
Wooden plaque—148

X

Xerophiles—11

Z

Zygomycotina—20, 21
Zygospore—21

★★★

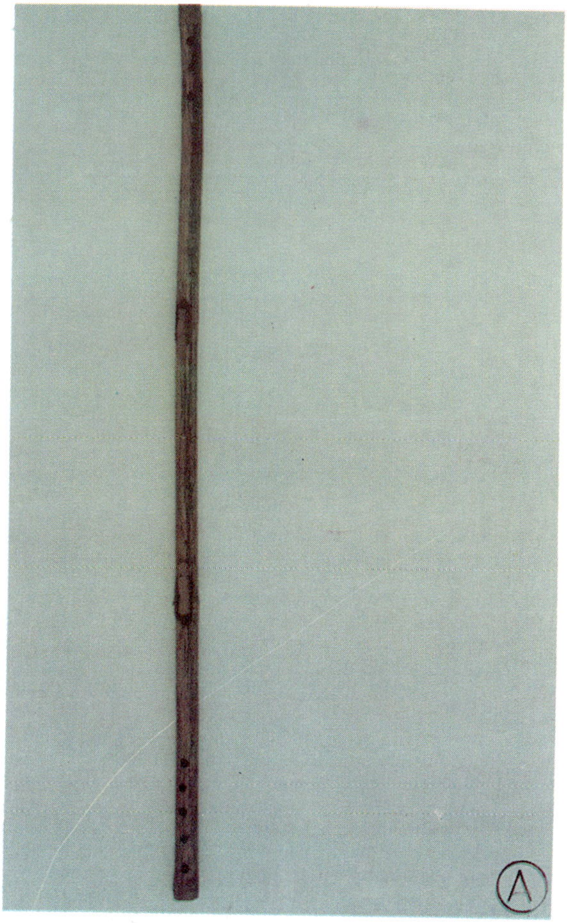

Pl. I A. Photograph of the Flute, a musical instrument made up of bamboo, preserved in the Dept. of Museology, M.S. University, Vadodara.

Pl. I B. Pavri, an important musical instrument used by different tribals. The black coloured substance as well as fruit body harboured the fungus.

Pl. I C. Wooden Chinese screen displayed in Museum and Picture Gallery, Baroda. It is decorated with precious stones.

Pl. I D. Fungus affected pages of an illustrated Parsi manuscript book displayed in the Museum of the Dept. of Museology.

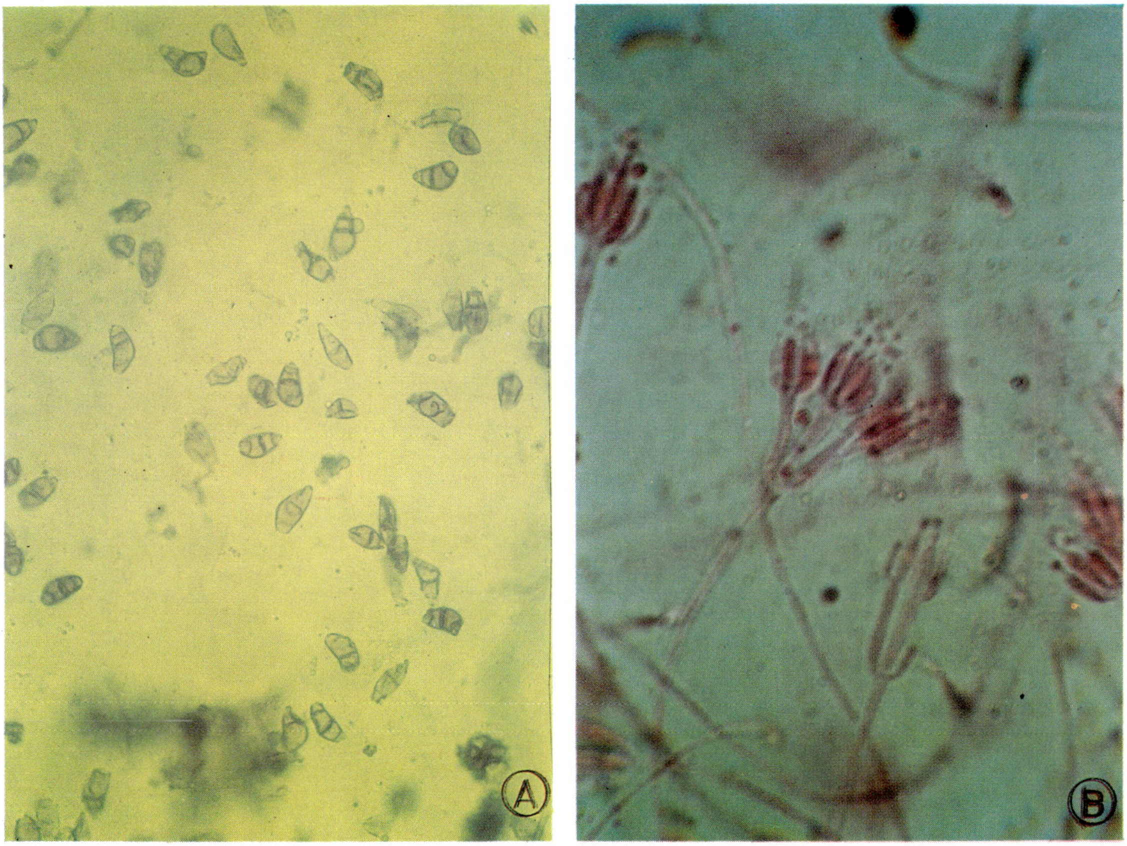

Pl. II A. Conidia of *Cochliobolus verrucuosus (Curvularia* state) obtained from Flute. X 126.

Pl. II B. Conidiophore and conidia of *Penicillium citrinum* obtained from Pavri. X 126.

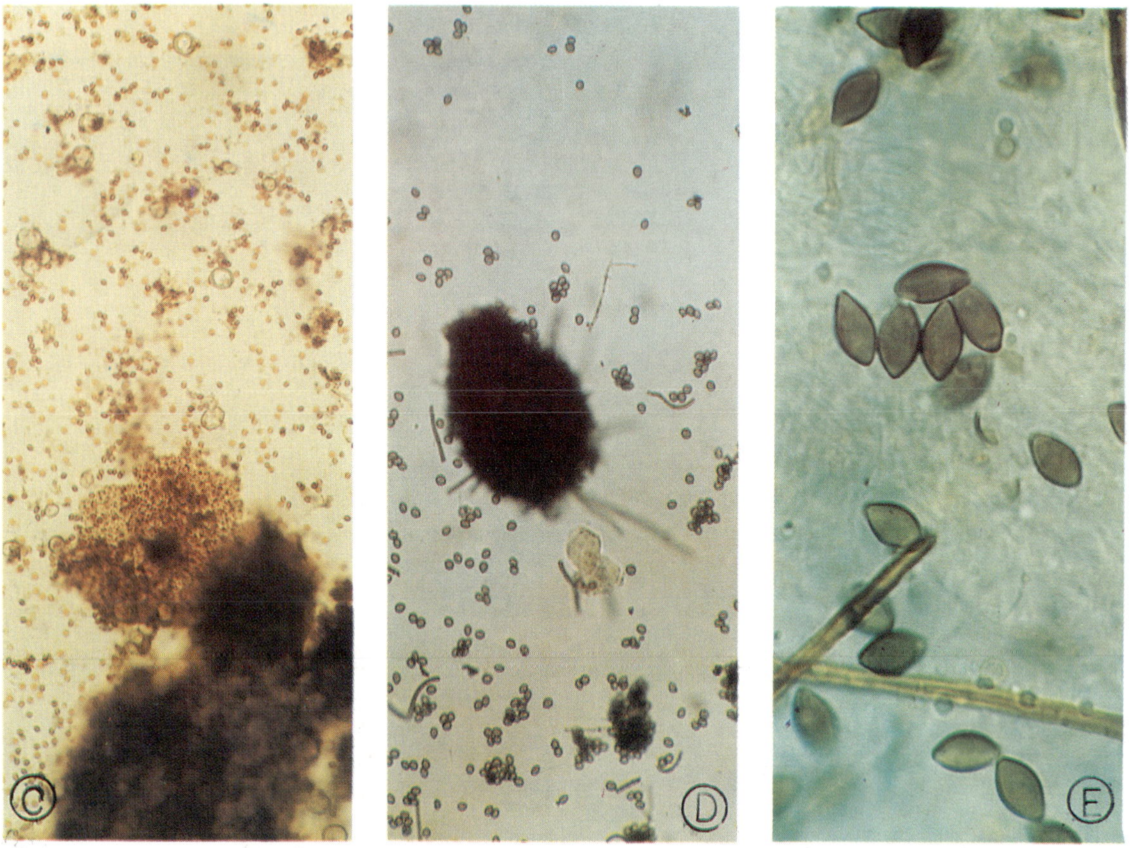

Pl. II C. Conidia of *Aspergillus ustus* causing damage to wooden Chinese screen. X 126.

Pl. II D. Perithecium of *Chaetomium cymbiforme* (Note the coiled hairs on perithecium) causing deterioration to the Parsi manuscript book. X 94.

Pl. II E. Enlarged ascospores of *Chaetomium cymbiforme*. X 126.

Pl. III A. Photograph showing Museum and Picture Gallery, Baroda situated in Sayaji Park, Baroda.

Pl. III B. Indian Art Gallery of Museum and Picture Gallery, Baroda where Indoor aeromycoflora was studied.

Pl. III C. Inner view of Ethnology gallery of Museum and Picture Gallery, Baroda where indoor aeromyco-flora was studied.

Pl. IV A. External view of the Department of Museology, M. S. University, Baroda.

Pl. IV B. Inside the Museum of the Department.

Pl. V A. Outer view of Smt. Hansa Mehta Library, M. S. University, Baroda.

Pl. V B. Books kept inside the library.

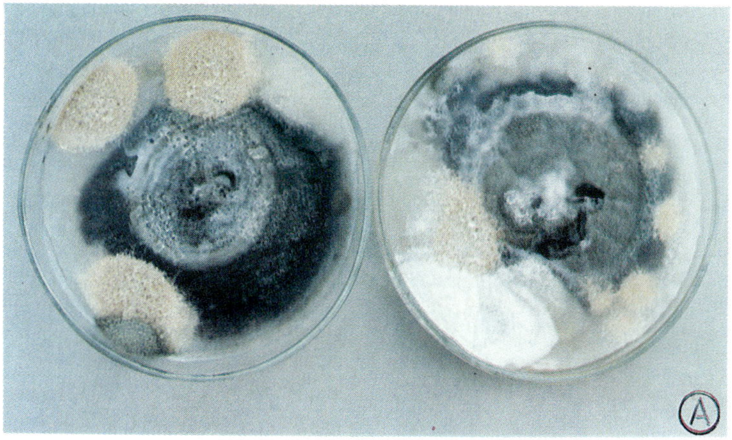

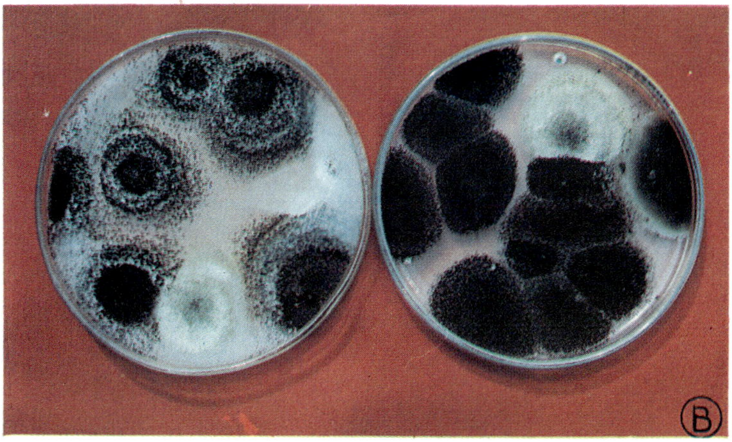

Pl. VI A. Petridishes showing fungal colonies in the experiment of indoor aeromycoflora of Museum and Picture Gallery, Baroda.

Pl. VI B. Petridishes showing fungal flora of the Museum of the Dept. of Museology, M. S. University, Baroda.

Pl. VI C. Colonies of fungi isolated from the air of Smt. Hansa Mehta Library, M. S. University, Baroda.

Pl. VII A. Induced growth of *Cochliobolus verruculosus* on the painting made on card board.

Pl. VII B. Induced growth of *Penicillium citrinum* on the painting made on card board.

Pl. VII C. Induced growth of *Aspergillus ustus* on the painting made on card board.

Pl. VIII A. Experiments on deterioration of Teak Wood in culture medium by *C. verruculosus* and *P. citrinum*.

Pl. VIII B. Experiments on deterioration of Deodar wood in culture medium by *C. verruculosus* and *P. citrinum*.

Pl. VIII C. Erlenmeyer flasks showing deterioration of filter paper *in vitro* by *C. cymbiforme*.

Pl. VIII D. Erlenmeyer flasks showing deterioration of common paper *in vitro* by *C. cymbiforme*.

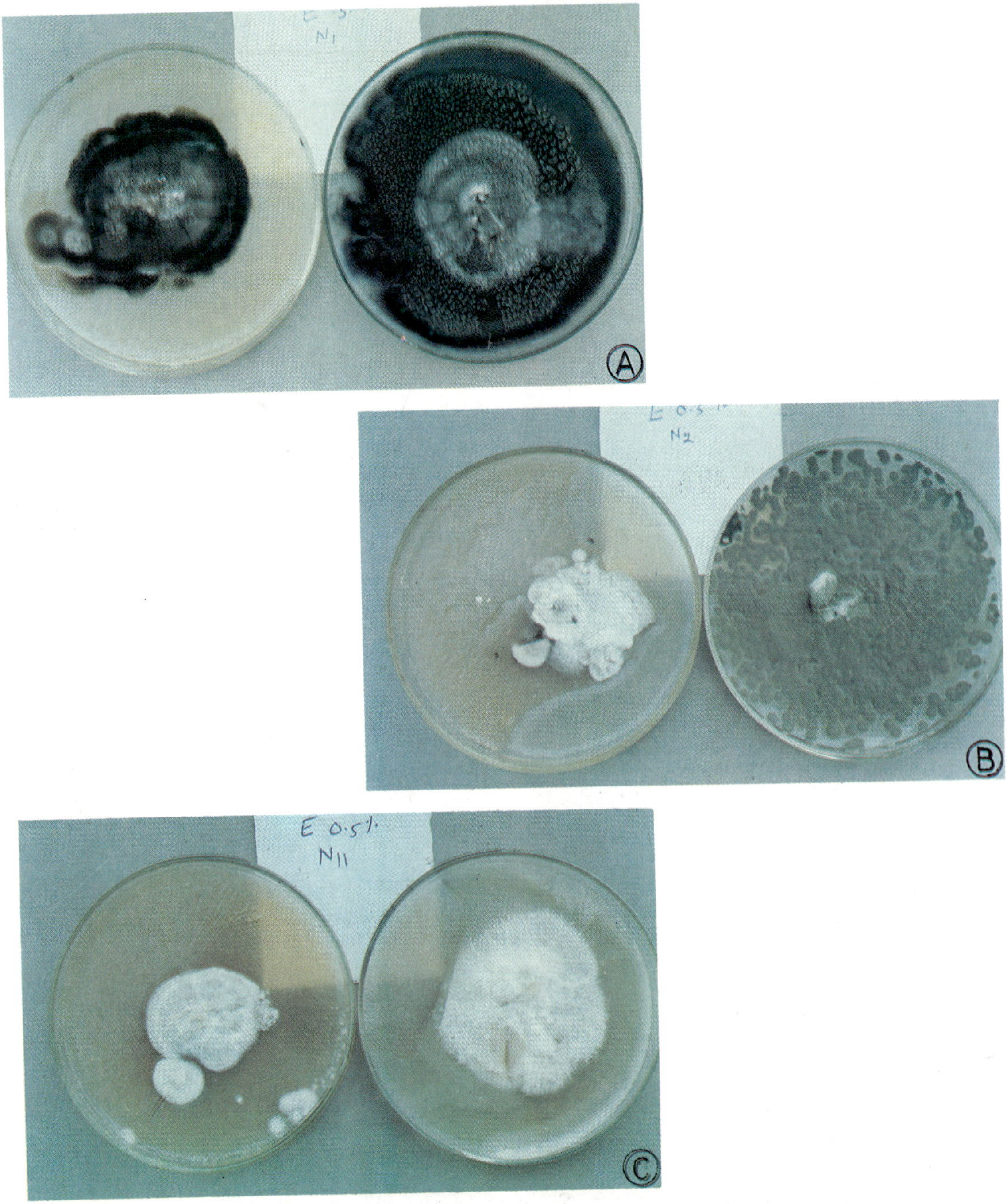

Pl. IX A. Effect of 0.5% Eucalyptus oil on *C. verruculosus* in control experiment.

Pl. IX B. Effect of 0.5% Eucalyptus oil on *P. citrinum* in control experiment.

Pl. IX C. Effect of 0.5% Eucalyptus oil on *A. ustus* in control experiment.